HYDROLOGY
AND
WETLAND
CONSERVATION

A Volume in the *Water Science Series*

Published on behalf of the Institute of Hydrology

HYDROLOGY
AND
WETLAND
CONSERVATION

Kevin Gilman
Institute of Hydrology

JOHN WILEY & SONS
Chichester • New York • Brisbane • Toronto • Singapore

Copyright © 1994 by Institute of Hydrology,
Maclean Building, Crowmarsh
Gifford, Wallingford,
Oxon, OX10 8BB, England

Published by John Wiley & Sons Ltd,
Baffins Lane, Chichester,
West Sussex PO19 1UD, England
National   Chichester (0243) 779777
International   +44 243 779777

Other Wiley Editorial Offices

John Wiley & Sons, Inc., 605 Third Avenue,
New York, NY 10158-0012, USA

Jacaranda Wiley Ltd, 33 Park Road, Milton,
Queensland 4064, Australia

John Wiley & Sons (Canada) Ltd, 22 Worcester Road,
Rexdale, Ontario M9W 1L1, Canada

John Wiley & Sons (SEA) Pte Ltd, 37 Jalan Pemimpin #05-04,
Block B, Union Industrial Building, Singapore 2057

Library of Congress Cataloging-in-Publication Data

A catalog record for this book is available from the Library of Congress

British Library Cataloguing in Publication Data

A catalogue record for this book is available from the British Library

ISBN 0-471-95152-8 √

Camera-ready copy supplied by the Institute of Hydrology
Printed and bound in Great Britain by Bookcraft (Bath) Ltd

Contents

Preface

The management of wetlands for conservation, alongside land used for agriculture or development, requires an understanding of the principles of wetland hydrology and of those hydrological processes which have maintained the site's natural habitats in the long term. The author has examined the hydrological behaviour of a range of wetlands in the UK and from this sample have been drawn some general conclusions which can be used in the assessment of wetland sites. Development of all kinds can pose a threat: although dewatering has an obvious, immediate and lasting impact, agricultural drainage is by no means the only potentially damaging operation. Analysis of threatened wetland sites observed over a period of 17 years shows that mineral extraction, housing development and road-building are also significant.

This study outlines the major freshwater wetland types represented in the UK, from marsh, which could be regarded as a pioneer community taking over from shallow open water, to bog or acid mire, which represents the final divorce of the wetland plant community from its mineral substrate. A selection of seven sites, distinguished by unusually long or intensive sets of hydrological data, has been used as a basis for the scientific investigation of wetland hydrology. The analysis of observed water levels from dipwell networks and continuous water level recorders and their seasonal and temporal variation, led to the development of the wetland lysimeter. This new instrument can be used to carry out detailed and automated measurements for the determination of the specific yield, actual daily transpiration and net lateral groundwater flow.

The wetland water balance differs from that of terrestrial sites in the dominance of storage of water in the ground. This persists as a controlling factor throughout the history of the site. Saturated conditions in the soil, and the low velocity of water movement, lead to a chemically reduced root zone, to the formation of peat and to the development of specialised plant communities. These special conditions underlie the conservation value of wetland sites as unusual natural habitats which were often protected from human interference in the past by the impracticability of improvement using small-scale manual methods.

Hydrometry on wetlands necessarily concentrates on water level measurement, both in the soil and in open water bodies. The seasonal variation of water levels observed on wetland sites reflects the pattern of precipitation input and evaporative demand. Short-term fluctuations, for instance the effects of rainfall and the daily cycle of evaporation, are superimposed on this seasonal variation.

The ground surface of peatlands or mires moves vertically in response to the amount of stored water as it often consists of a shallow mat of vegetation supported by a considerable depth of loose peat or organic mud. The mechanism of ground movement was investigated at two sites, Crymlyn Bog, Swansea, and West Sedgemoor, Somerset. Clear evidence exists that the dewatering of the upper layers in the peat profile causes a reduction in buoyancy forces and a consequent seasonal cycle linked closely to the cycle of water level change. Ground movement influences both the implementation of hydrometric measurements on peatlands and modifies the effects of dewatering on the plant community.

The storage of water in the saturated zone is characterised by one central parameter, the specific yield, which is a function of the porosity and hence of the degree of compaction of the substrate. Peat exhibits a wide range of variation in its hydraulic properties according to its history and its position in the soil profile. Specific yield controls the extent of water level change in response to the annual climatic cycle and to transient phenomena such as droughts and short-term dewatering. A simple method for determining the specific yield from rainfall inputs was produced successfully; this can be applied to a continuous record of water level using daily rainfall measurements.

Given an estimate of the specific yield, diurnal fluctuations in a continuous groundwater level record can be used to estimate daily actual transpiration from a wetland community. In a mixed fen or "litter" community at Wicken Fen, Cambridgeshire, this method demonstrates that actual transpiration, which in summer makes up a large proportion of evaporation, is always less than the potential evaporation. There was a clear indication of the significance of the cutting of the "litter" community at two-yearly intervals: the largely green vegetation that grew up after cutting transpired more rapidly than the older vegetation, which included much standing dead material.

Many wetlands are bounded or crossed by open ditches, which according to land use and/or season can be considered either as drains or subirrigation channels. The interaction between these ditches and the adjacent soil is of particular importance and dipwell transect data from three sites have been presented to show the relationship between ditch water levels and groundwater levels over the year. The permeability of peat is generally low and the effect of a ditch on groundwater levels is limited at distances exceeding about 50 m. The annual cycle of groundwater levels can be predicted quite accurately using a mathematical model and such a model, taking account of spatial variation in the hydraulic parameters, was used to simulate water levels at West Sedgemoor.

This improved estimates of specific yield and permeability and explained the unusually rapid attenuation of ditch effects with distance.

While it is possible to use solely water level measurements, supplemented by climate data, in the investigation of wetland hydrology, there are uncertainties introduced by the lateral flow of groundwater and the routing of rainfall to the water table. A lysimeter overcomes the problem of lateral flow by isolating a soil block from its surroundings, but lysimetry can be labour-intensive. To overcome this, an automated system was devised which could form the basis for the routine measurement of transpiration in wetland communities, including tall fen vegetation. The lysimeter has been demonstrated at a bog site in mid-Wales and at West Sedgemoor, where it gives results which agree well with estimates of actual evaporation made by several independent methods. The results from the un-grazed bog site, where the standing crop contains dead material and transpiration starts quite late in the season, appear to corroborate those from Wicken Fen, while West Sedgemoor demonstrates the effect of grazing and mowing in rejuvenating the crop and increasing transpiration rates. The results also show inhibition of transpiration on days with very high potential evaporation rates.

Finally, attention is drawn to two clear findings from this work:

●   The influence of open drains on the water table in adjacent ground is very limited in extent in UK wetlands, especially in fine silt and peat. On such slowly permeable soils, rainfall during the winter, in the absence of significant evaporation, is adequate to raise the water table until it is at or near the ground surface, and within each field groundwater flow is not rapid enough, in the absence of field drainage, to cause a significant lowering of the water table outside a narrow peripheral strip. What lowering does occur over the summer within the field areas is caused primarily by evaporative demand. This is not to say that ditching cannot have a deleterious effect on a wetland site. There is evidence to suggest that the maintenance of a relatively low water level in ditches over the winter does prevent or reduce winter flooding, for instance at Wicken Fen and West Sedgemoor, and obviously the character of a site with standing water could be seriously changed by ditches controlled in that manner.

●   Contrary to many indications contained in the literature, actual evaporation from UK wetlands during the summer, when transpiration dominates, appears to be lower than the potential rate, particularly where natural or semi-natural communities comprise a significant proportion of dead material. It is probable that literature results are biased towards swamp communities and many reported experiments may have been subject to the "oasis effect", which enhances evaporation rates from lysimeters improperly exposed. It is difficult to carry out a systematic exploration of the subject through published papers, which have often been contradictory, but the results of this study from three contrasting communities (tall fen / intermediate-height sedges and rushes / managed grassland), suggest that wetlands, especially where they carry tall vegetation, have a positive contribution to make to water resources.

Somerset Levels and Moors. In summer, the ditches or rhynes are kept at a high water level to provide 'wet fencing'. This practice also helps to compensate for evaporative demand from the fields, which produce a substantial hay crop, even in dry years.

Somerset Levels and Moors. An extreme network of arterial drains prevents or at least reduces the impact of flooding of this low-lying area.

Skewbridge Bog

Cors Erddreiniog

CHAPTER 1

Introduction

The countryside of Europe is a mosaic of differing and often conflicting land uses that vies in its complexity with the natural habitats that preceded the arrival of agricultural man. The improvement of land for increased agricultural production, probably dating in northern Europe from the elm decline in Neolithic times at around 3000 BC (Moore & Webb, 1978), is an ancient activity and of unquestionable value to society. Now, however, the balance has changed and agriculture is no longer a practice carried out in scattered clearings in the wildwood, nor is there a stable pattern of land use hallowed by long tradition. Modern agriculture is an industry which, like any other, seeks to maximise its profits to expand and increase its efficiency: attempts to steer it away from a course of progressively increased production have yet to be shown to be effective (MAFF, 1989) and in many parts of the country there is conflict between farming and natural habitats.

The threat of improvement of the remaining areas of natural habitat, combined with pressures from every other sort of development ranging from mineral extraction to housing and roads, has roused powerful forces in favour of conservation. In the UK the designation by the Nature Conservancy Council (NCC) and its successor bodies English Nature, Scottish Natural Heritage (SNH) and the Countryside Council for Wales (CCW) of important areas as Sites of Special Scientific Interest (SSSIs) has helped to formalise the process of deciding the future of significant remnants of a range of habitats. A relatively new factor is the commitment of large sums of money by non-governmental conservation bodies to land purchase as a means of protecting sites.

Some natural habitats can co-exist with agricultural activity provided the latter is not too intensive. For instance, moorland and montane grasslands have long been a valued component of upland stock farms and semi-natural woodland is prized for timber production and to provide shelter for stock and game. Wetlands, however, have always been a prime target for reclamation. Their value for grazing is very limited in the natural state and agricultural productivity can be raised dramatically by drainage, which also allows better access for ploughing, pasture re-seeding and fertiliser application or even for arable crop production. The result is that in the late twentieth century large tracts of wetland, for instance the mosses of north-western England and the Fens of East Anglia, have been virtually turned over to intensive farming with relatively small remnant areas of natural habitat. Marginal, poorly-drained land abutting on smaller wetland areas has been drained and improved. Land values of wetland sites are low and other development pressures, diverted from built-up areas and high grade agricultural land come to bear fully on the wetlands. Wetland sites are widely acknowledged to be valuable reservoirs of wild species, exhibiting considerable diversity and this has been recognised by a programme of mapping, recording and designation, yet it is difficult if not impossible to find a wetland site of any significant size in the UK that is not either under direct pressure for development or subject to threat from activities on its periphery.

It is instructive to examine the nature of the pressures upon wetlands. Over a period of 17 years, the author has been consulted on the hydrology of a large number of wetland sites in the UK. Of the 59 studies which have led to detailed reports ten were directly or indirectly affected by agricultural drainage and several others could be said to be affected by changes in the nutrient budget as a consequence of agricultural activities on adjacent land. On 11 sites advice was required on positive water management to maintain or improve the conservation value. The remainder of the sites were affected by development of other kinds, notably mineral extraction, road construction and the abstraction of water from surface and subsurface sources. Figure 1 summarises the pressures on the 48 threatened sites.

Although the development pressure on wetland sites is not always agricultural, it is dewatering by drainage of various kinds that constitutes the greatest single indirect threat. Drainage of adjoining farm land is carried out by a network of field drains, usually mole drains and pipes, leading into a grid of open ditches. Frequently a ditch forms the boundary of a remnant wetland site. Other developments, for instance the extraction of peat, gravel or opencast coal or the construction of roads, have a similar requirement for efficient dewatering and hence a similar effect on the water table

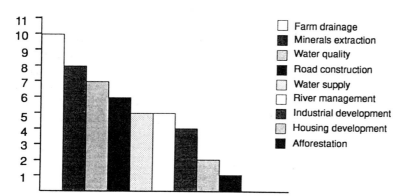

Farm drainage
Minerals extraction
Water quality
Road construction
Water supply
River management
Industrial development
Housing development
Afforestation

FIGURE 1 Of a sample of 48 wetland sites investigated 10 were affected by agricultural drainage: there are many other pressures on UK wetlands.

within the adjacent wetland site. The boundary between natural habitat and developed land becomes not just the demarcation between two land uses, but the buffer between areas of very different hydrology. If the effects of drainage are not to propagate further than they were ever intended, there must be a clear answer to the superficially simple question of how to reconcile the interests of nature conservation and agriculture on the two banks of the same ditch. The answer to the question cannot be so simply expressed. Only active water management in the wetland area can prevent slow degradation of the wetland as a consequence of water level lowering in the interests of development while any such management must not have deleterious effects on the adjoining land. It is also the case that the water needs of crops and the techniques for water management on agricultural land, for instance drainage and irrigation, are better understood than are the demands of the diverse wetland community which has evolved to exploit quite subtle distinctions in water quality and availability.

Active hydrological management of a wetland reserve must have an established basis of scientific measurement if it is to be effective. Even in those sites whose future as nature reserves has been thought for many years to be assured by sympathetic ownership, hydrological data are not generally available. Concern about a decline in water level is often based only on anecdotal evidence, or on observed floristic changes, which could arise from causes unrelated to water as, for example, from changes in grazing or other management practices.

This book investigates those important aspects of wetland hydrology that have to be taken into account for the purpose of water management. It draws on data collected over a network of sites representing the main wetland types, bringing out similarities in the response of wetland sites to the climate and other influences. The analysis of data from wetland sites, mainly in the form of groundwater level measurements, can yield important insights into the likely response of the sites to changes in their surroundings.

Field data collection is a time-consuming and expensive occupation, particularly if its aim is to monitor the response of a system to changes in natural climatic variables, for example rainfall events or the cycle of evaporation. The climate in the UK, even in the summer, rarely cooperates with planned environmental investigations. An instrumental approach is described here, using recent developments in microcomputer technology to carry out semi-controlled experiments in the field. Data collected on a continuous basis by a wetland lysimeter can improve our understanding of the detailed hydrology of wetland sites, offering better design of water management schemes and more confident and accurate prediction of the effects of hydrological changes.

CHAPTER 2

Freshwater wetlands in the UK

THE RANGE OF WETLAND TYPES

Wetlands are distinguished from other terrestrial habitats by having a significant excess of water for a large proportion of the time. This excess water imposes an important control on the natural vegetation. Through its effects on the soil atmosphere, on the chemistry of the soil and on the range of plant species that can compete successfully under the special conditions created by slowly moving or stagnant waters, it continues to have an effect throughout the development of a site. Although the dividing line is not clear, it is suggested that continually flooded ground should only be counted as wetland if the water is relatively shallow and vegetation is of the emergent type.

The source of the excess water may be high rainfall, regular inundation by floodwaters as on valley-bottom sites adjacent to large rivers, or groundwater emerging as springs. Wetlands also develop by natural succession on the margins of open-water bodies, by the buildup of silt or organic debris. Wetlands are often created by the impedance of natural drainage, as in closed depressions, on terrain modified by glacial or periglacial processes in valleys with very low gradients or on impermeable soils.

Wetlands are rarely static habitats: through the accumulation of mineral or organic matter, a wetland site generally (but not invariably) moves in the direction of drying out as the ground surface rises relative to the water table. This process of terrestrialisation or *Verlandung* goes hand in hand with the natural succession of the wetland vegetation communities. Only in the later stages, as the habitat tends towards climax woodland, or management imposes a semi-natural vegetation community, for instance a wet pasture or meadow, does a wetland site have the appearance of having entirely forgotten its origins. Evidence from peat bogs shows that the process may undergo reversal even at a late stage as the climate changes and mire vegetation again becomes established. The Boreal period between 9500 and 7000 BP (years before present) and the sub-Boreal period between 5000 and 2500 BP were characterised by the spread of pine forest over peat bogs, only to be overtaken by renewed peat growth in the wetter Atlantic and sub-Atlantic periods between 7000 and 5000 BP and between 2500 BP and the present (Lowe & Walker, 1984).

Wetland sites fall into three main types, bogs, fens and marshes, which are distinguished on the grounds of the presence of peat and on the base status of the water. Bogs and fens, the peatland type, are grouped under the term mire. Peat, which forms the soil of bogs and fens, is an organic material which develops from vegetable debris whose breakdown by soil bacteria and other decomposers is prevented, usually by the absence of oxygen under conditions of saturation with stagnant or very slowly moving water. Other factors which encourage peat development are low temperatures, acidity and shortages of important nutrients. The buildup of peat deposits on a site is equivalent to the long-term storage of a significant proportion of the organic production of the ecosystem in a form that is protected from the normal processes of decomposition.

Peat development is not an inevitable feature of the wetland habitat. If the movement of water in the soil is sufficient, the process of humification is more rapid, large decomposers such as worms, insect larvae and other arthropods may thrive and the soil remains largely mineral. The term "marsh" is reserved for these mineral-based wetlands, which are usually in receipt of a significant input of alluvial sediment and are flushed by moving water.

The velocity with which water moves horizontally through or over the soil is an important factor determining the type of vegetation community. Rapid flow, provided that erosion is prevented, may give rise to a preponderance of taller species and a greater overall production of plant material (Sparling, 1966). Flushing by moving water helps to carry away toxic substances, for example ferrous iron, manganese and aluminium, and it can provide a sufficient flow of valuable nutrients to the roots even if concentrations in the water are low.

Wetland type	Principal features
Marsh	Mineral soils, but build-up of organic substances more rapid than in terrestrial environment Regular inundation with surface water Water movement and/or nutrient input prevents peat development Vegetation grassy or herb-rich
Fen	Peat, developing slowly. Nutrient-rich and base-rich waters. Vegetation grassy or herb-rich.
Bog	Peat developing quite rapidly (1-2 mm y^{-1}) Acidic, nutrient-poor and base-poor waters Vegetation contains significant mossy component

TABLE 1   The three main wetland types

At many wetland sites there is a clear progression with time from open water or marsh to fen to bog, the later stages being dependent on local conditions. In very active environments, in which the water flow is rapid and there is a ready supply of silt, the system does not develop beyond marsh. Even so, the less active corners of a site, for example cut-off river channels and areas protected by levees, may provide the opportunity for waters to clear, floating rafts of emergent vegetation to spread and, by producing peat and organic muds, to progress to fen.

Lowering of the water table, or the slow process of peat buildup, may allow the growth of scrub and the succession to a carr, or fen woodland, community (Figure 2). Local variations in topography, minor hummocks rising slightly above the surrounding base-rich waters, lead to an increase in the importance of rainfall as a water supply and provide the opportunity for bog-mosses to grow. These islands of a more acidic environment, which are frequently *emersive*,

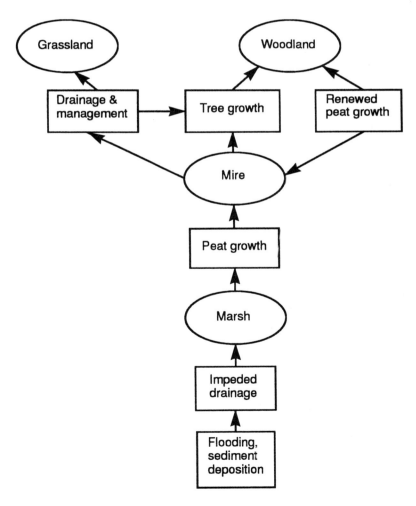

FIGURE 2   Processes in the development of wetlands

i.e. able to move vertically, floating on a mat of peat and undecomposed vegetation, may act as nuclei for the spread of bog communities. The phenomenon of ground movement or *Mooratmung*, which is an important process tending towards the protection of wetlands from water level changes, will be discussed in a later chapter.

The succession of wetlands from marsh to bog was traced by Walker (1970) for a large number of sites in the UK, drawing on pollen diagrams and stratigraphy. Although the transitions from open water to mire did not follow an inflexible pattern, it was clear that reedswamp, a marsh community dominated by the common reed *Phragmites australis*, had occurred at an early stage in more than half the sites studied. The arrival of *Sphagnum* mosses, whether in reedswamp, fen or carr appeared to herald the transition to a bog community.

The marsh stage is represented in the deepest layers of the substrate of many mire sites as an early phase, immediately following the last glaciation. In this phase there was a copious supply of weathered mineral material providing waters with a high base status. Sites occupying closed depressions have deposits of fine lacustrine clays, indicating an input of sediment-laden runoff from higher ground, only partially filtered by marginal vegetation. In marsh systems, owing partly to the oxygenation by moving water, there is little opportunity for the preservation of plant remains and the substrate builds up by deposition of mineral material.

Marsh communities are found today over a more limited range in the UK, usually in the more active environments where water movement or the deposition of silt impede the succession to mire. As in late-Devensian and pre-Boreal times (12000 to 9500 BP), marsh species are also frequent colonisers of shallow open water bodies, for example ponds and gravel pits. The characteristic marsh plants are the emergent species which form a marginal community in the shallows of open water bodies and can persist in drier habitats where inundation, though infrequent, remains a significant factor. The ubiquitous *Phragmites australis* (common reed) is the best-known member of this group, other common dominants being *Typha latifolia* (reedmace), *Glyceria maxima* (reed sweet-grass) and *Iris pseudacorus* (yellow iris). Slightly drier conditions along the margins of open water bodies are exploited by the sedges, for example *Carex elata* (tufted sedge), *C. paniculata* (greater tussock sedge) and *C. riparia* (greater pond sedge) and by the reed canary-grass *Phalaris arundinacea*. Meadows which are periodically flooded, usually in winter and early spring, tend to be humus-rich and carry, in addition to a wide range of grasses including the prominent tussocks of *Deschampsia cespitosa* (tufted hair-grass) and *Molinia caerulea* (purple moor-grass), a variety of herbaceous plants such as the familiar marsh marigold *Caltha palustris*, the ragged robin *Lychnis flos-cuculi*, the purple loosestrife *Lythrum salicaria* and the great willowherb *Epilobium hirsutum*.

Lake clays, and silts which may have been the substrate for marsh habitats, are often overlain by peat rich in the remains of *Phragmites australis* (common reed), which can advance as a floating mat across shallow open water, eventually creating a fen habitat. In fens there is little mineral matter at the surface, and water movement is moderately slow, but the vegetation community must be attuned to the presence of base-rich water. Cors Erddreiniog, Anglesey, a calcareous fen, shows a sequence of development in which lake clays were superseded by a finely-divided calcareous marl deposit when the growth of vegetation on high ground and a marginal swamp community had reduced the input of fine sediment to the lake. The most recent deposit, which covers the entire site, giving way to a deep bed of marl only in the littoral zone of the modern lake, is fen peat formed from the remains of tall grassy vegetation such as *Phragmites australis* (common reed) (Figure 3).

Fen vegetation generally consists of tall grasses and sedges with a tendency towards the formation of monodominant stands which have been exploited by traditional rural industries concerned for example with the provision of reed for thatching. Peat forms slowly from fallen leaf litter and dead stems, but flooding with nutrient-rich waters prevents the succession to more acid communities except in localised "islands" where a slight elevation provides a degree of isolation from the standing water and groundwater. Meade & Blackstock (1988) describe the occurrence of acidophilous islands at Cors Goch, Anglesey, surrounded by vegetation that is typical of lime-rich conditions.

The key to the transition from marsh to fen appears to be the exclusion of alluvial silt, perhaps by the buildup of natural levees between river and backswamp, as at Llangloffan Fen, Dyfed, and the Black Fens of East Anglia (Godwin, 1978) or by a reduction in the sediment load of tributary streams, as may have happened on an extensive scale in the late-Devensian period.

Eventually the buildup of peat over a fen site results in a rise in the ground surface relative to the water table and isolation of the

FIGURE 3 Stratigraphic transect across Cors Erddreiniog, Anglesey. The transect crossed only the northern tip of the modern lake, Llyn yr Wyth Eidion, which is up to 8 m deep near its centre. The fen peat is the substrate that most influences the present hydrology of the site, though it is probable that the relatively firm marl and clay prevents ground movement which in a deeper peat body might tend to counteract drainage effects.

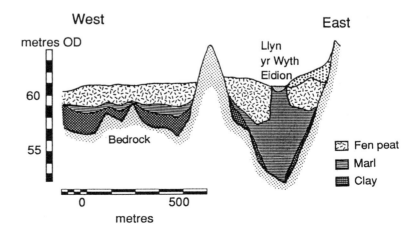

ground surface from the base-rich groundwater will lead to the development of carr (Godwin, 1936) or bog communities (Walker, 1970). Alternatively, agricultural development may remove the fen from the wetland spectrum by deep drainage and use for arable crops, horticulture or forestry, or encourage a succession to wet grassland by a moderate degree of drainage.

Management of fen communities is an important factor in their conservation. This was not generally appreciated in the early days of the establishment of nature reserves and on Wicken Fen, for instance, it was many years before the practices of the moribund sedge-cutting industry, which had maintained an open-fen community for centuries, were re-instituted in the interests of habitat restoration and conservation (Godwin, 1978). By that time a large area of the fen had succeeded to carr.

The true peat bog is a late stage in wetland development in which the surface vegetation has become chemically divorced from the mineral soil by the buildup of a sufficient depth of peat. The nutrient input to the bog is almost all from direct precipitation, supplemented by minor inputs from dry deposition, for instance of airborne dust and sea salts, from the atmosphere. Many bog sites are domed with a drainage pattern that is radially outward. These raised mires range in size from hummocks a few metres across to large bogs occupying whole floodplains as at Cors Caron, Dyfed (Tregaron Bog). The input of water and nutrients is entirely from the atmosphere and such mires are described as ombrogenous, a term which draws attention to direct rainfall as the source of water, and *oligotrophic*, owing to their low nutrient and base status. Closely related to the ombrogenous mire, but with their nutrient budgets supplemented to a moderate degree by runoff and drainage of nutrient-poor waters from surrounding ground, are soligenous mires and poor fens, frequently found in upland valleys in the west of Great Britain.

One of the most striking features of bog communities is the detailed structure that is revealed by careful study: the bog surface has an intricate topography on a variety of scales, ranging from microrelief elements on a scale of metres (ridges, hummocks, pools and hollows) to larger areas containing a mosaic of microrelief elements. Typically an undisturbed bog consists of several quite clearly defined regions in which the microrelief elements take a similar form. Ivanov (1975) classified these regions as *microtopes* and developed a detailed system of mapping which could be used for survey of extensive mires by aerial photography. The microrelief of the bog is linked with the vegetation community in a complex system of feedback loops. Certain species (for instance the hummock-forming mosses *Sphagnum fuscum* and *S. imbricatum* and the species confined to waterlogged pools and hollows, which include *S. cuspidatum*) are capable of surviving only a very limited range of water table variation and the diversity of the vegetation community is inseparably linked with the diversity of the microrelief. While the mosses make up a substantial proportion of the bog community, other plant groups are present, their distribution also connected with microrelief. The common cotton-grass *Eriophorum angustifolium*, a range of low-growing sedges such as *Carex limosa* (bog sedge) and *Rhynchospora alba* (white beak sedge) and the insectivorous sundews *Drosera anglica* (great sundew) *and D. intermedia* (oblong-leaved sundew) all occupy the floors of shallow pools and hollows. The hummocks provide a niche for a range of dwarf shrubs such as the cross-leaved heath *Erica tetralix*, the bog myrtle *Myrica gale* and the cranberry *Vaccinium oxycoccus*, for the bog asphodel *Narthecium ossifragum* and the round-leaved sundew *Drosera rotundifolia*.

SAMPLE SITES USED IN THIS STUDY

The choice of sites for scientific study is always difficult: no small group of sites is going to be representative of the full range of freshwater wetlands in the UK or even of those in which hydrology and conservation are important issues and the usual problems of access, availability of background information and length of record apply. The selection listed below, though certainly not representative, includes sites in which there has been a conflict between agricultural activity and conservation (Cors Erddreiniog, West Sedgemoor and less directly Cors-y-Llyn, Llangloffan Fen and Wicken Fen) and a site at which industrial activity has been the chief threat (Crymlyn Bog). At six of the sites, water management is important for future conservation and, in most cases thanks to the dedication of wardens and their staff, there is a record of water levels extending over several years. The remaining site, Skew Bridge Bog, Llangurig, Powys, is a small soligenous bog which was conveniently situated to serve as a test-bed for mire lysimetry.

Cors Erddreiniog, Anglesey

The fens of Anglesey, Cors Erddreiniog, Cors Goch, Cors-y-Farl and Cors Bodeilio, derive their botanical interest from their location in the west of Britain, where a relatively high rainfall is combined with climate and geochemical influences which are related to the proximity of the sea. The Carboniferous limestone (Greenly, 1919) and gentle topography of this part of Anglesey have given rise to a number of mires of the 'rich fen' type, at least three occupying former lakebeds, with a diverse flora showing a wide range of pH tolerance.

On the Anglesey Fens, plant species representative of sites as far apart as the fens of Northumbria and East Anglia are growing together. The sites support a mixture of northern and southern floristic elements: *Carex elata* (tufted sedge), *Cladium mariscus*, the orchids *Dactylorchis traunsteineri* and *Ophrys insectifera* (fly orchid) are in the main southern species, while *Carex lasiocarpa* (downy-fruited sedge) and the orchid *Dactylorhiza purpurella* occur in more northerly parts of Britain (Ratcliffe, 1977).

At Cors Erddreiniog (SH 470830), the supply of carbonate-rich groundwater, which enters mainly as springs and streams from the limestone escarpment along the eastern edge of the site, is counteracted on a local scale by water from the contiguous sandstone and boulder clay and by rainwater, so that the slight elevation of hummocks and tussocks above the general groundwater level provides more acidic conditions. There are areas dominated by *Phragmites australis* (common reed), *Cladium mariscus* (giant saw-sedge), *Molinia caerulea* (purple moor-grass), *Schoenus nigricans* (black bog-rush) and *Juncus subnodulosus* (blunt-flowered rush). *Sphagnum* mosses are present in places where more acid conditions have developed (Thomas, 1976).

Cors Erddreiniog is partly owned by the Countryside Council for Wales (CCW, formerly the Nature Conservancy Council, NCC) and is designated as a National Nature Reserve (NNR) within a larger area scheduled as an SSSI. The fen occupies a shallow lake basin and the peat, which varies in depth from 1.5 to 4.5 m, is underlain by a plastic clay of late-Devensian and pre-Boreal age (this appears to correlate with similar clays found by Seddon, 1957, at the nearby Cors Goch) and by a finely-divided carbonate marl. A small but deep elliptical lake, Llyn yr Wyth Eidion, is all that remains of the once-extensive area of open water.

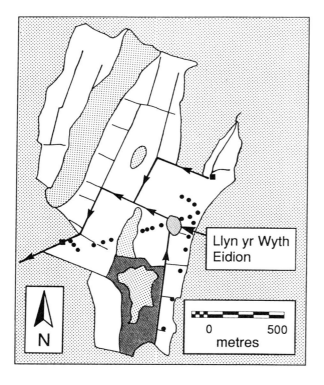

Llyn yr Wyth Eidion

0   500
metres

N

Figure 4 Cors Erddreiniog, Anglesey. The network of dipwells is indicated by circles: the two continuous water level recorders are marked by black squares. Arrows show the direction of flow in the main ditches. The cross-section in Figure 3 follows the line of the main east–west transect of dipwells.

Conservation problems at Cors Erddreiniog stem from attempts to improve the arterial drainage of the site in the 19th and 20th centuries. A reticular pattern of open drains dissects the site into a number of approximately rectangular compartments. Although a detailed ground survey commissioned by the NCC showed that wastage of the fen peat had occurred along the lines of all significant ditches, the main ecological effect of drainage was the deflection of the natural succession towards dominance of *Molinia caerulea*, the purple moor-grass, and until recently there was no pressure to back up the arterial network with field drains in the interests of further improvement. A change in land ownership led to a conflict between the intensification of grazing on part of the fen, supported by field drainage and more frequent ditch maintenance, and the interests of the NCC, which centre on the conservation of the diverse fen community.

Arguments about the influence of ditch water levels on the native flora were supported by the results from an extensive network of 19 dipwells installed by the Institute of Hydrology (IH) on behalf of the NCC in 1979 and 1980. Readings of the water table level between 1979 and the present have shown the seasonal pattern and have indicated longer-term changes resulting from water level management within the NNR (Figure 5).

Although an extensive network of dipwells can show the seasonal cycle of water levels, indicate when changes due to interference or management have occurred and help to explain the distribution of vegetation communities, groundwater levels can vary considerably on a local scale, particularly in response to ditch water levels. A linear transect of dipwells, 10 at intervals of 2 m, was established in 1979, and extended in 1980, adjacent and perpendicular to the main outlet ditch. The results of observations over a number of years suggest that the zone of influence of the ditch on groundwater levels is very narrow: discussion later in this volume will attempt to

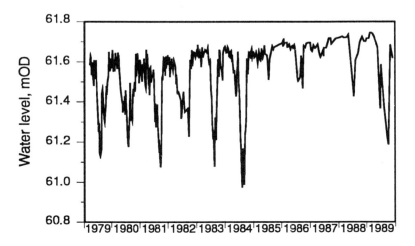

FIGURE 5 Groundwater levels at dw 3, Cors Erddreiniog, 1979-1989. Dipwell 3 is in a Molinia-dominated community which appears to have replaced true fen as a result of 19th century drainage (Meade & Blackstock, 1988).

compare these observations with the NCC survey, which showed that one important effect, peat wastage, propagated much further into the mire expanse.

Cors-y-Llyn, Powys

Cors-y-Llyn (SO 016553) is an elliptical basin mire situated on a plateau to the east of the River Wye near Newbridge, Powys (Figure 6). Although Llyn is a small site, it has considerable conservation value deriving from the fact that it has been subject to very little human interference and it preserves a semi-natural vegetation community ranging from acid bog to peripheral birch carr. The deep peat profile has preserved a valuable record of the

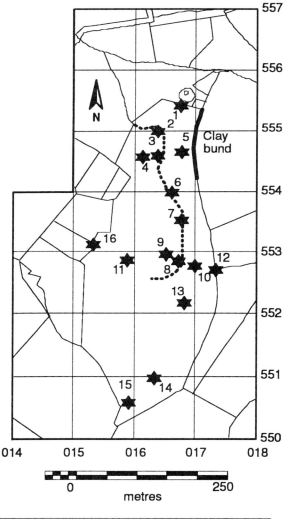

FIGURE 6 *Cors-y-Llyn, Powys.* The dipwell network is arranged to follow a boardwalk which protects the sensitive mire expanse from trampling. Off-boardwalk dipwells are read remotely. Continuous water level recorders are mounted on dipwells 3 and 8.

vegetational history of the mire and the surrounding area (Moore & Beckett, 1971). The mire occupies a basin on the axis of a small catchment on the Wenlock shales which are generally considered to be impermeable: there is no evidence as yet to indicate the presence of a significant groundwater input to the mire.

Moore & Beckett (1971) carried out a floristic classification on the results of 277 quadrats, dispersed across the site on a regular grid. Species groups appeared to divide the mire into four distinct sub-areas: two elliptical mire expanses to the north and south of a crescentic strip which carried tall pine trees, and a narrow peripheral birch carr or lagg fen which almost surrounded the site. The peripheral carr, which indicates more base-rich waters draining from surrounding pasture land, appears to have been a persistent feature of the site, at one time having been rather more extensive, covering the northern basin. Peat cutting has modified the natural succession and brought about a change to the present mire community which is based on the colonisation of cut-over areas in the north by *Sphagnum recurvum*. Pine woodland is a recent feature and small trees which had spread over the mire expanse have been removed recently by the NCC wardens.

Drainage of peripheral land, restricted to shallow grips and ditches along hedge boundaries, has always gone on in the catchment of the mire but concern was expressed in the early 1980s about drainage close to the single outlet from Cors-y-Llyn. A weir was installed to exercise a partial control over water levels in the peripheral carr and work was started on a clay-cored bund to sever ditches which threatened to convey water more rapidly away from the mire expanse.

Evidence relating to the decline of water levels at Cors-y-Llyn is partly anecdotal: it was believed that the hummock-hollow systems on the mire expanse were not as wet as before and that the 1 m thick layer of liquid peat recorded by Moore and Beckett (1971) had disappeared. No water level data at all were available in 1983.

The NCC wisely acknowledged the need for hydrological observations to test the effectiveness of the water management schemes. Two continuous water level recorders were installed in 1983, one in each of the two main mire expanses, and these records were supplemented in early 1985 by a network of 16 dipwells whose water levels were measured fortnightly. All water level readings, the water levels being measured from fixed metal datum posts inserted to the base of the basin, have continued up to the present. Figures 7 and 8 show seasonal changes in water level over the observation period. Changes in water level resulting from management, though detectable, were quite subtle when compared with the annual range of variation. The similarity between the two Figures is evidence that the seasonal variation is adequately expressed by readings taken at fortnightly intervals but it is clear that the degree of detail in the continuous record will allow a much closer study of day-to-day changes in the water balance.

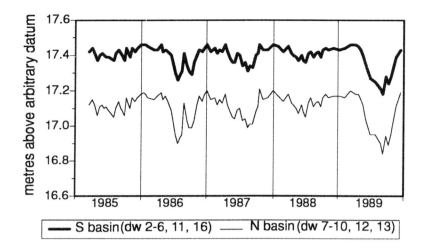

Figure 7 Groundwater levels recorded at Cors-y-Llyn, 1985 to 1989. Llyn may be divided into two basins: the lines in the figure represent the mean groundwater level in each basin, computed from the periodic observations of dipwells in the two basins.

Crymlyn Bog, Swansea

Crymlyn Bog (SS 695945) is a large complex of bog and fen lying between the valleys of the Rivers Tawe and Neath and covering a total area of 268 ha. The mire expanse is only a few metres above sea level and slopes very gently from north to south. The southern margin of the Bog is formed by the sand dunes of Crymlyn Burrows which have been heavily developed for a trunk road, railways, a factory and an oil storage depot (Figure 9).

The solid geology of the Crymlyn catchment is dominated by the Pennant Sandstone unit of the Coal Measures which compose the east and west ridges (Institute of Geological Sciences, 1972). The lower slopes of the ridges are mantled with till and it is likely that the base of the Crymlyn valley is partly filled with this material and with outwashed and soliflucted materials derived from it. At the

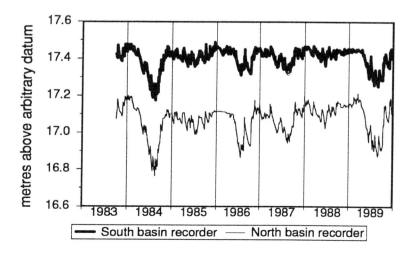

Figure 8 Variations in groundwater level at Cors-y-Llyn, as shown by the continuous water level recorders which were mounted close to dw 3 and dw 8. The plots here are of midnight values abstracted from the paper charts using a digitising tablet. During the summer there is a diurnal fluctuation in groundwater level of amplitude comparable with the daily fall in level. This diurnal fluctuation, which is caused by transpiration, will be the subject of a detailed discussion in Chapters 3 & 4.

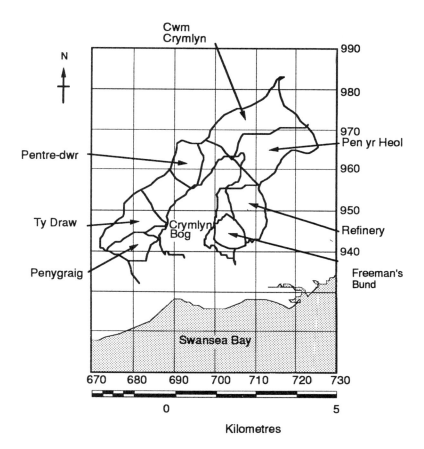

FIGURE 9 Crymlyn Bog, Swansea. The mire receives inflows from a number of small catchments around its periphery, including polluted waters from mine drainage and the oil refinery.

seaward end of the site there are interleaved deposits of dune sand and compressed peat, extending well below sea level, and Godwin (1940) used the records from boreholes penetrating these deposits to show that the sea level had advanced and retreated several times during the Boreal period, the total rise in sea level having been around 23 m. During the installation of dipwells by IH staff, between 3.7 m and 5.4 m of peat was recorded but one station near the seaward tip of the site had 10.4 m of soft deposits, probably peat or organic mud.

Industrial development of the catchment has left its mark on Crymlyn Bog: mine drainage inflows have changed the water quality locally and a disused canal runs almost the length of the site, while the southwestern part is affected by a municipal tip on the site of a now-demolished power station and its associated fuel ash tip. Along the eastern ridge, the oil refinery at Llandarcy has discharged oil-polluted drainage water to the Bog.

As with many wetland sites, there was a complete absence of information on water levels at a time when consideration had to be

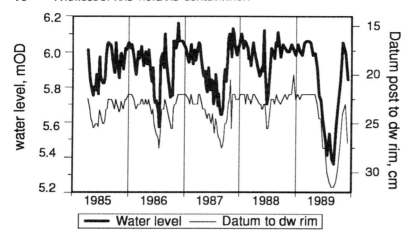

FIGURE 10 Variations in groundwater level recorded in dipwell dw 5 at Crymlyn Bog, 1985 to 1989. Measurements of the distance between the rim of the dipwell and the top of the datum post show that the dipwell and the ground surface rose and fell with the groundwater level. The ratio of ground movement and water level change for various stations in the dipwell network was between 5% and 12%.

given by the NCC to schemes for water management. In response to the need for data, a hydrological study by IH was commissioned. A network of 9 shallow dipwells was installed in 1985. The considerable depth of deposits at Crymlyn, and their fluid nature, raised the question of ground movement in response to groundwater level changes. Each dipwell was accompanied by a steel tube which penetrated the peat and was hammered firmly into the underlying silts to act as a fixed datum relative to which dipwell movements and water level variations could be determined. Regular readings of dipwell water levels were carried out from 1985 to 1989 (Figure 10).

Llangloffan Fen, Dyfed

Llangloffan Fen (SM 910318) is a valley fen occupying the floor of the valley of the Western Cleddau, or Cleddau Wen. The river formerly followed a meandering course but it has recently been canalised and deepened and land north of the river has been reclaimed by under-drainage to provide grazing.

The solid geology of Llangloffan Fen is dominated by relatively impermeable Ordovician rocks, shales of the upper Arenig and Llanvirn series dissected by glacial melt-water channels of which the Cleddau channel is one. Glacial drift is believed to be important in sustaining the many springs feeding the Fen, and lime-rich sands and gravels may be the source of the water supply for a stand of the calcicolous *Cladium mariscus* (greater saw-sedge).

Tall fen and marsh vegetation now occupies a strip of land south of the stream, about 1.4 km in length and 200 m wide. The vegetation community ranges from marsh, dominated by *Phalaris arundinacea* (reed canary-grass) close to the stream, through *Phragmites australis* (common reed) fen and willow and birch carr to sycamore and oak woodland and wet grassland dominated by *Molinia caerulea* (purple moor-grass) at the foot of the valley flank. A stand of *Cladium mariscus* (greater saw-sedge) and an area of *Sphagnum* moss hummocks with *Vaccinium oxycoccus* (cranberry) are present near the edge of the carr.

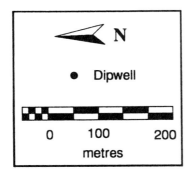

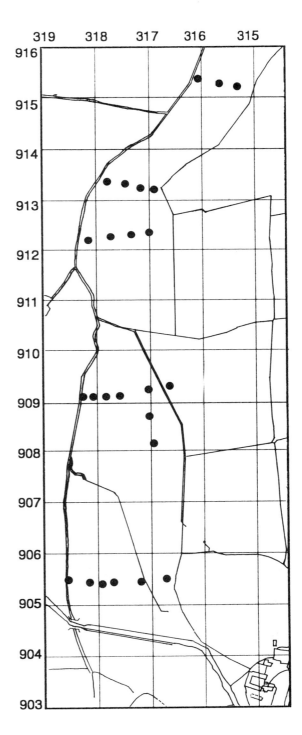

FIGURE 11 Llangloffan Fen, Dyfed. The Western Cleddau flows eastwards along the northern edge of the site, and dipwell transects sample the range of communities from stream-side marsh through fen and carr.

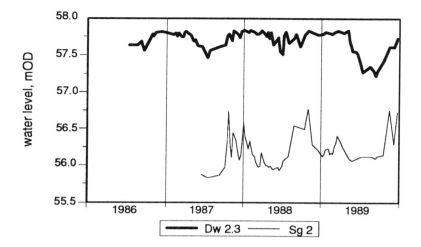

FIGURE 12 Variations in groundwater level at Llangloffan Fen. Dipwell 2.3 is in fen peat and is most strongly influenced by the annual cycle of the water balance. Water level readings at staff gauge sg 2 on the stream about 70 m north of the dipwell are also plotted and show little correlation with peat groundwater levels.

Peat at Llangloffan occupies a narrow strip separated from the stream by a silt levee or roddon. Fen vegetation in the backswamp behind this roddon has led to the deposition of up to 2 m of peat and at one location the peat depth is in excess of 5 m.

On the recommendation of IH, eight dipwells, arranged in two transects, were installed by NCC staff in 1986, four more in 1987 and 11 more in 1989. A continuous water level recorder has been in operation on one of the wells since 1986. Seasonal variations of water level in the peat show that the annual cycle of rainfall and transpiration has more effect on groundwater levels than does the stream water level, but the combination of the higher permeability of peat close to the surface and the intersection of artificial and natural drainage lines by the deepened stream channel has probably led to a decline in groundwater levels.

Skew Bridge Bog, Llangurig, Powys

Development work on the mire lysimeter (Chapter 4) demanded easy access to a wetland site on a daily basis and the Llangurig mire offered a convenient site within a few miles of the IH Plynlimon station. Skew Bridge Bog (SN 925800) is a small remnant of a larger expanse of soligenous mire (nutrient-poor fen) occupying an interfluve site in the valley of the River Wye. It is crossed by a disused railway line and land to the north of the line has been drained to provide grazing. Drainage of the area to the south, by gripping, has been unsuccessful and the mire remains wet for most of the year (Figure 13).

Waters entering the mire from the high ground to the south are acidic and the mire carries a community dominated by *Molinia caerulea* (purple moor-grass) in sloping areas, with the common

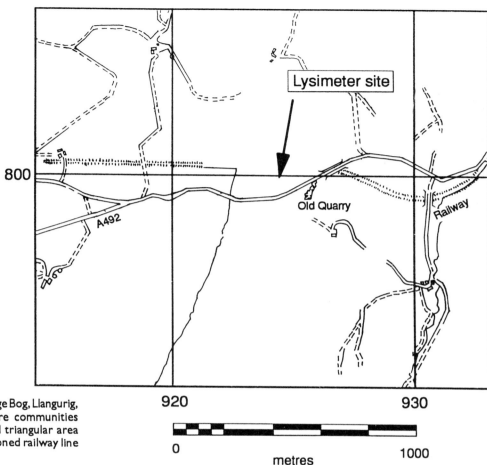

FIGURE 13 Skew Bridge Bog, Llangurig, Powys. Natural mire communities now occupy a small triangular area between the abandoned railway line and the road.

cotton-grass *Eriophorum angustifolium*, and the downy-fruited sedge *Carex lasiocarpa* present in the wetter area occupied by the lysimeter site.

West Sedgemoor, Somerset

West Sedgemoor (ST 370320) is a drained fen that fills the flat floor of a northeasterly-facing valley on the southern edge of the Somerset Levels. The peatlands or "moors" of Somerset occupy a similar position to the Black Fens of East Anglia, being protected from the sea by the broad band of silt "levels" that extends from the Mendips in the north to the Quantocks in the south. West Sedgemoor has an extensive highland catchment formed by Triassic mudstones including the Keuper Marl, capped in places by Jurassic clays of the Lower Lias group (Institute of Geological Sciences, 1969, 1975). Water entering from the highland catchment at the southwestern extremity is led through a broad canal, the West Sedgemoor Main Drain, along the northwestern flank of the Moor to the West Sedgemoor Pumping Station, where it is raised using diesel and electric pumps into the River Parrett.

In complete contrast to the other sites listed in this report, West Sedgemoor is an area subject to moderately intensive agricultural use. Its vegetation, even on fields that have never been subjected to pumped drainage, may at best be considered semi-natural, but the site is of immense value to nesting and feeding birds. Sedge- and herb-rich grassland is harvested for hay and grazed: since notification as SSSI in 1982 farming operations have been subject to the normal consultation procedures which require advance notification of potentially damaging operations such as changes to the drainage and cropping routine. In addition to the Main Drain, there is an extensive reticular network of around 100 km of narrower drains, locally known as "rhynes", which divide the Moor into a large number of rectangular fields (Figure 14). Traditionally, the rhyne network is held at a high water level over the summer, if necessary by allowing water through sluices from the Parrett. This provides "wet fencing" for cattle grazed on the Moor after the hay crop is taken. In winter, the Main Drain is pumped to keep the rhyne system at a low level to provide flood storage, thus reducing the depth and duration of inundation by waters from the catchment.

This traditional pattern of water management, which arguably was a sustainable and sympathetic use of the peatlands, was upset in the late 1970s by agricultural development in the form of localised pumped drainage schemes. Subsequent events at West Sedgemoor, some of them melodramatic, have led to increased ownership by conservation bodies, notably the Royal Society for the Protection of Birds (RSPB) and eventually to a vigorous and continuous debate on the interactions between farming and conservation in the whole of the Somerset Levels and Moors.

A long sequence of measurements of water levels measured in the Main Drain at the West Sedgemoor Pumping Station was insufficient to characterise the hydrological behaviour of the Moor. In 1985 IH began a study to quantify the hydrological components of the water balance of the Moor, investigating the past operation of the West Sedgemoor Pumping Station and the relationship between the nominal penning level and actual levels in the rhyne system and monitoring the movement of groundwater within the fields in response to climate and rhyne levels. This work subsequently developed into a full investigation of the water use of the Moor.

Many of the findings of the West Sedgemoor studies, which were reported to NCC and NRA-Wessex (Marshall & Gilman 1989, Gilman, Marshall & Dixon 1990, Gilman & Marshall, 1991), are outside the scope of this volume. However, the groundwater data from West Sedgemoor were used in the development of a model which is described in outline in Chapter 3 and the dipwell network installed on West Sedgemoor, four transects of five dipwells

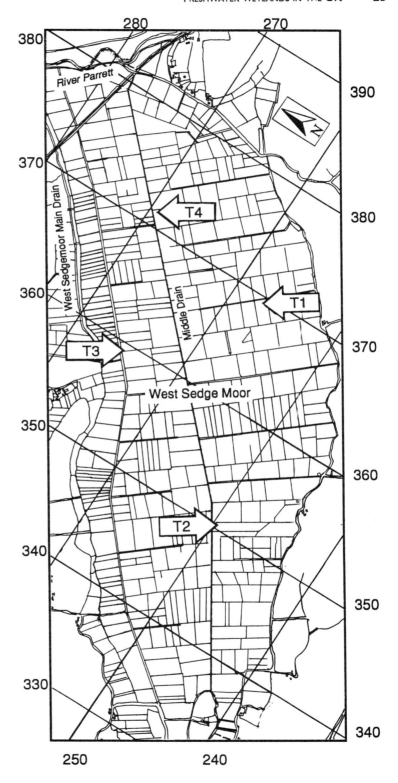

FIGURE 14 West Sedgemoor, Somerset. The mire lysimeter was installed adjacent to dipwell transect T4. Grid squares are kilometres.

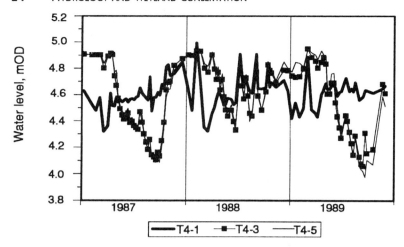

FIGURE 15 Variations in water level on dipwell transect T4 at West Sedgemoor, 1987 to 1989. Dipwell T4-1 is 2 m from the edge of the rhyne, T4-3 22 m and T4-5 42 m from the rhyne. The seasonal pattern of water levels in the rhyne, frequently disrupted by flood events, is supplanted in the centre of the field by a regime controlled by the annual cycle of rainfall and evaporation.

perpendicular to rhynes, provided a convenient background against which to run the mire lysimeter over the summer of 1990. Furthermore, the phenomenon of ground movement was investigated by a number of experiments at West Sedgemoor. Figure 15, a plot of groundwater levels in dipwell transect T4, should be sufficient to highlight similarities in the movement of the water table between this site and the others in the list.

Wicken Fen, Cambridgeshire

Wicken Fen (TL 552703), which consists of 245 ha of calcareous fenland, represents a remnant of a once-extensive landscape, that of the Black Fens of East Anglia. Though it is doubtful that Wicken was ever typical of the Black Fens, even before the General Drainage that was completed in 1663 (Rowell, 1983), having been maintained for the production of "sedge" (*Cladium mariscus*, greater saw-sedge)

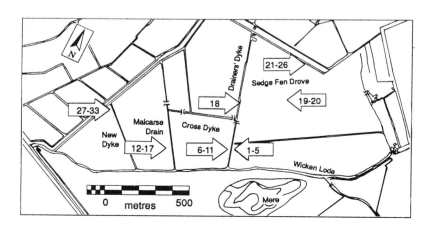

FIGURE 16 Wicken Fen, Cambridge-shire. The positions of dipwell transects are shown by arrows. The majority of dipwells were set out in straight-line transects extending from major dykes. Two wells, 19 and 20, were fitted with continuous water level recorders.

from 1600 or before, its modern value lies in the fact of its survival as one of the few remaining undrained portions. The fen is sustained by the Wicken Lode, a navigable high-level drainage channel running from Wicken village to the lock at Upware.

The peat of Wicken is underlain by the impermeable Gault clay which has provided core material for flood embankments and was once the foundation of a thriving brick industry at Wicken (Evans, 1925; Institute of Geological Sciences, 1974). The peat is probably between 3 and 5 m deep and in places it contains a thin layer of shell marl close to the surface which indicates the presence of a persistent flooded area in the past (Hanson, 1976). Records of peat digging at Wicken, though difficult to interpret, appear to suggest that it was confined to the area farthest west and the area immediately adjacent to the Lode near Wicken village (Rowell, 1983).

The preservation of Wicken Fen, against a background of inventive and determined efforts to drain and convert the Fens to intensive agricultural and horticultural use, has resulted partly from the dedication of naturalists who have passed over the ownership of land, parcel by parcel, to the National Trust and partly from its position alongside one of the high-level lodes that are a feature of the South Level of the Fens. The major lodes, Bottisham, Swaffham, Reach (with tributaries to Burwell and Wicken) and Soham, were Chalk streams impounded and embanked to provide navigable channels to the villages on the edge of the Fens and to prevent flooding by waters from the higher ground to the southeast (Farren 1926). The Wicken Fen reserve consists of three sections: Wicken Sedge Fen which is the accessible area lying to the north of the Wicken Lode and two areas of lower ground to the south of the Lode, Adventurers' Fen and St. Edmund's Fen.

At Wicken, a long period of cropping for sedge for thatching and "litter" (a mixed herbaceous community containing common reed *Phragmites australis* and purple moor-grass *Molinia caerulea*) for animal bedding had created a broad expanse of open fen vegetation which has largely succeeded to carr in the years since the decline of cutting. The history of the exploitation of Wicken Fen for peat, sedge and litter has been discussed in detail by Rowell (1983). Today a significant area (18 ha of litter and 10.5 ha of sedge out of a total of 134 ha on Wicken Sedge Fen) is maintained as fen by cutting, a labour-intensive operation, which is now justified on conservation grounds rather than as an economic activity.

Despite early and detailed hydrological investigations over the years 1928-30 (Godwin, 1931; Godwin & Bharucha, 1932), new management proposals advanced in the early 1980s were without the support of the archive of hydrological information that could have been built up if the early work had been followed up on a routine basis. Though Gowing (1977) had repeated some of Godwin's observations and concluded that the Fen had been affected by a lowering of the Lode water level in the intervening years, it was difficult to relate the later readings of groundwater levels with

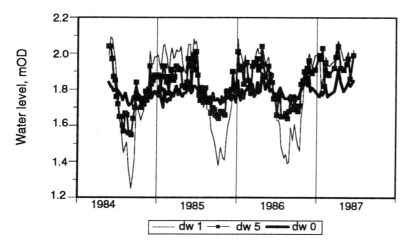

FIGURE 17 Variations in groundwater level for the dipwell transect east of Drainer's Dyke, Wicken Fen, 1984 to 1987. The dipwells were set 10 m apart along a line perpendicular to the Dyke: dw 1 at 50 m and dw 5 at 10 m. dw 0 was excavated by Gowing in 1977, and is hydraulically connected with the Dyke.

Godwin's. A new network of 33 dipwells, distributed across the Fen, was installed by IH in 1984 and water levels were read fortnightly (weekly for the more accessible sites) by the Trust warden and his team between May 1984 and 1987. The dipwell network was backed up by two continuous water level recorders on small fields cut for litter. Figure 17 shows the water level record for selected dipwells on one of the dipwell transects.

CHAPTER 3

The wetland water balance

The effective management of a wetland area depends on a thorough understanding of the way in which the components of the water budget (Figure 18) interact to provide a stable, though seasonally varying, template against which wetland plant communities can develop. The plant species present, and their spatial distribution, are dependent on a range of environmental variables and on the ability of the various species to compete, grow to maturity and reproduce. Godwin (1931) recognised that the elevation of the water table, the principal controlling environmental variable, could exercise an influence through winter flooding and the varying seasonal resistance of plant species to deficient aeration or through low summer levels leading to a water shortage in the upper soil. The control exerted by excess water is not direct: its influence is mainly through the reduction in oxygen availability in the root zone and the development of chemically reduced conditions to which many wetland species, but few dry land species, are adapted (Gosselink

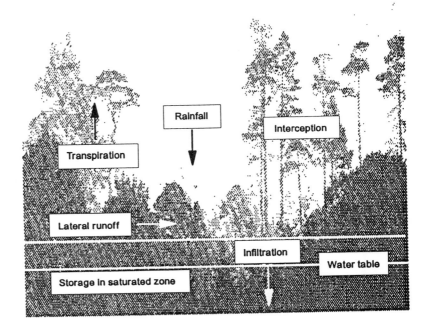

FIGURE 18 Components of the water budget of a wetland site. The precipitation input, in the UK mostly in the form of rainfall, is partitioned into interception loss and net rainfall, and further loss to the atmosphere occurs by transpiration from plants and evaporation from bare soil and open water. Surface flow and lateral groundwater flow may be regarded as throughputs, but at at any time, a given area of wetland may be receiving a net input from these lateral flows, or contributing a net output to areas downstream.

& Turner, 1978). Lateral flows, by providing an influx of sediment, nutrients and bases and by controlling the range of variation of the water table or water level, are important determinands of the wetland habitat.

Certain elements of the water budget, for instance surface runoff and groundwater seepage, are open to artificial control in the interests of conservation but other components, being climatically influenced, can introduce extremes which are beyond human control. In this chapter the consideration of the natural seasonal and short-term changes in the most easily monitored component, groundwater storage, leads on to a discussion of the various components of the water budget in the context of wetlands and the fundamental hydrological concept, the water balance.

WETLAND HYDROMETRY

Hydrometric networks in the UK are based on the notion of the catchment or drainage basin which is the fundamental unit of water resources. The National Rivers Authority operates a country-wide network of river gauging stations, the general distribution of which is chosen according to the needs for flood warning and the assessment of resources, while precise location is dictated by practical considerations. Raingauges and climate stations, though strongly influenced by historical factors and the availability of observers willing to take daily readings, give a good coverage of the UK but are often found lacking when local data is required. Groundwater observation wells are also sited according to resources considerations, usually on large-scale aquifers, and are of limited use in assessing the hydrology of wetland areas.

The needs of wetland hydrology are usually so local in scale that the long-established national stations for groundwater level and streamflow are at too great a distance to be of use and represent an integration of processes over a large area, only a small proportion of which is wetland. The hydrology of a wetland site can be assessed only on the basis of local data which usually have to be collected for the purpose. The simplest data to obtain, and frequently the only data available as time series, are water levels.

Water level measurement

Wetland hydrometry differs from catchment hydrometry in that greater emphasis is placed on the measurement of stored water, either in open water bodies, or as groundwater under water table conditions. Flows in slowly-moving ditches or through the soil are difficult or impossible to measure directly and the determination of evaporation proves as challenging on a wetland site as on any other.

Those components of the water budget that involve movement to and from storage, particularly rainfall and evaporation, induce measurable responses in the variation of the groundwater level and the level, or stage, of standing and flowing water. These responses can be used in the evaluation of the two most difficult components to measure directly, the actual evaporation and the net groundwater flow. When combined with the techniques of digital modelling, studies based on close observation of water level changes can provide a powerful means of assessing the hydrological behaviour of a wetland site.

Open water

Water levels in open water bodies are conveniently measured using a simple graduated staff gauge, securely mounted on a post anchored to a firm substrate. Although the idea is simple, the problems are legion, ranging from vandalism to the effects of ice-heave and the difficulty of finding firm ground close to open water on wetland sites. Repeated survey of the staff gauge zero is important and the provision of a secondary datum point nearby may be worthwhile. For instance at Crymlyn Bog a staff gauge that was clearly visible from a footpath was supplemented by an unobtrusive metal post which could act as a datum from which to measure water level should the gauge be damaged. Periodic checks were made to detect any relative movement.

The use of continuous water level recorders on open water bodies requires the installation of a rather more secure mounting, an instrument shelter and a stilling well to eliminate the effects of waves on the float. Nevertheless, on flowing channels where the water level varies on a short timescale, the continuous record provided by an automatic instrument has considerable value, though the analysis of paper charts is very labour-intensive. The move away from chart recorders towards digital data loggers promises a more cost-effective method of collecting and processing data at short time intervals.

Water in the soil

Storage of water in the soil is monitored by the measurement of the elevation of the water table. The water table is not the upper limit of the zone of saturation as is sometimes asserted. It is the level at which the water pressure is equal to the atmospheric pressure and hence is the level at which water will stand in a well that is hydraulically connected with the groundwater body. Dipwells for observation networks are excavated, often by hand, to a depth that is below the expected lower limit of variation of the water table and in wetland sites rarely exceed 2 m in depth. To prevent sloughing of the hole, and to provide a datum for water level measurement, dipwells are cased and screened, usually with perforated or slotted plastic tubing but occasionally with other materials such as drainage tile (Burke, 1961) and they should be capped to prevent ingress of small animals, stones and other debris and to avoid unrepresentative readings caused by recent rainfall or drip from trees.

On some sites there is a significant vertical movement of water within the saturated zone. This can arise where there is upward flow from springs within the wetland, in peripheral areas of raised mires and adjacent to drainage ditches where hydraulic gradients are large. Flow of groundwater takes place in response to a hydraulic gradient and vertical hydraulic gradients can be measured by using piezometers in place of dipwells. Instead of an extensive perforated casing, a piezometer has a very localised screen to allow water movement and measures the hydraulic head at a point. Piezometers are usually installed in "nests", several piezometers, with screens open at various depths, occupying the same map location and usually inserted in the same drilled hole.

Because the response of the dipwell to changes in the water table demands the movement of water between the dipwell and the groundwater body, dipwell measurements can show a time-lag and this may be important when the water table is varying rapidly. To follow a rapid change in water table it is necessary to minimise the volume of inward and outward flow of water by using as narrow a dipwell as possible and this can cause problems in reading. In wide dipwells the water surface is visible and a steel tape can be used to measure the depth to water. In narrower wells, and where the depth to the water table exceeds about 1 m, an electric contact gauge is used. Electric contact gauges can be used in wells down to about 15 mm in diameter. Ingram (1983) gives a detailed account of other methods for measuring or indicating the water level, notably an air-bubbling method which can be used in very narrow wells and piezometers, and a pressure-bulb recorder which reduces or eliminates the time-lag.

All water level measuring stations at a site should be surveyed by levelling to a precision equal to that of water level measurement. For short-term studies the survey need not be related to Ordnance Datum but such is the scarcity of water level measurements from UK wetland sites that any sequence of data is likely to become of long-term significance, in which case the datum heights may assume great importance in the future. The minimum requirement is the setting up of permanent benchmarks on firm ground: examples are the datum point at Cors-y-Llyn, the earth anchors installed at Llangloffan Fen and the pillar set up by the Great Ouse River Authority at Wicken Fen.

The pattern of water level changes

Groundwater levels in wetlands exhibit a strong seasonal variation, which can be identified in data from a wide range of sites (Figures 5, 7, 8, 10, 12, 15, 17). Usually this takes the form of relatively high and constant water levels in winter and spring, followed by a decline in early summer to a minimum in late summer and a rise during autumn, reaching the winter level around mid-winter. The decline may be relieved by extreme rainfall amounts during the summer. This pattern, which matches broadly the variation in soil moisture deficit (SMD) computed from rainfall and evaporation estimates, was noted and analysed by Godwin (1931) in his studies of Wicken Fen, using a continuous water level recorder specially developed for the purpose. The pattern of water level variation was found to be strikingly similar in 1928 and 1929, in spite of differences in the temporal distribution of rainfall, and this demonstrated quite conclusively that the cause of the summer decline was evapotranspiration of the fen vegetation which is relatively conservative from year to year. In East Anglia transpiration is considerably in excess of rainfall in the summer: only in 1930, when the rainfall for July and August was almost double the long-term average for those months, was summer rainfall found to have an important effect on the fen water table. Godwin observed that the period of rapid fall in water level took place when the fen vegetation was growing most rapidly while the period of subsequent rise in the water table corresponded with the yellowing and dying-off of the fen vegetation.

Superimposed on the seasonal march of water levels are short-term fluctuations which can be attributed to rainfall events, the effects of changes in open water levels and the diurnal cycle of evaporation and transpiration. The immediate effect of rainfall is a rapid rise in the groundwater level: the decline that follows may be relatively rapid as lateral flow takes place over and near the ground surface or it may be simply the re-establishment of the decline caused by evaporative demand. Short-term changes in open water levels have an effect that falls away rapidly with distance. During dry periods, the day-to-day decline in groundwater level takes the form of a regular series of steps or saw-tooth variations as the diurnal cycle of evaporation withdraws water from the groundwater body during the daylight hours.

THE COMPONENTS OF THE WATER BALANCE

The concept of the water balance embodies the fundamental principle of the conservation of mass. The idea is no less valuable for being self-evident: in wetlands, for instance, it is the recognition of the obligatory balance between inputs, outputs and storage that leads to the investigation of sources of water and of the reasons behind the very existence of wetland outside of areas with high rainfall. An apparent imbalance has often led to the closer examination of hydrometric measurements and the recognition of hitherto undetected processes, and the presentation of a water balance is usually an essential part of hydrological investigations.

At the centre of the theory of hydrology is the catchment, a natural drainage basin defined by an obvious watershed. The topographic form of the catchment boundary is such that surface flow into a catchment area is zero and by careful choice of catchment area and geology groundwater flow can also be eliminated from the water balance equation. The flow out of the catchment can be measured by installing and operating a river gauging station at a suitable point. In a wetland area catchment boundaries are almost inevitably on high ground, boundaries which cross the wetland area itself, e.g. on interfluve mires, being diffuse and tricky to locate. Gauging of surface flows is difficult and it is often more appropriate to carry out the accounting procedure that is inherent in water balance calculations for a small elementary area in the wetland expanse than to use the wetland's periphery, i.e. its junction with higher ground, as a boundary. The water balance equation expresses the relationship between inputs, outputs and storage of water in the terrestrial phase of the hydrological cycle:

$$\text{Sum of inputs} = \text{Sum of outputs} + \text{Change in storage}$$
$$P_{net} + Q_{in} + G_{in} = E + Q_{out} + G_{out} + \Delta s$$

where P_{net} is the net precipitation, i.e. that portion of the precipitation that reaches the ground;

Q_{in} and Q_{out} are surface flows into and out of the elementary area;
G_{in} and G_{out} are groundwater flows into and out of the elementary area;
E is actual evaporation, the sum of transpiration and evaporation from bare soil and open water;
and Δs is the change in storage of water in the soil.

The lateral flows of surface water and groundwater possess both an incoming and an outgoing component: the elementary area receives a net lateral flow equal to the difference between inflow and outflow, so it is convenient to regard these flows as throughputs. Certain simplifying assumptions can be made: lateral flow in the unsaturated zone above the water table is very small in wetlands and it may be assumed that all unsaturated flow is vertical. There is generally little deep percolation in wetlands and groundwater flow, owing to the gentle hydraulic gradients, can usually be regarded as horizontal. Furthermore, there will be periods when either rainfall or evaporation, together with the change in storage, dominate the water balance, e.g. in the summer when water levels are below the ground surface and surface flow is zero. At night the evaporative demand falls to zero and the effects of lateral groundwater flow can be observed without the influence of evaporation.

Storage

Wetlands are distinguished by the presence of excess water exerting an influence on the climate, the properties of the soil and the range and distribution of plant species. Partly because of the gentle topographic and hydraulic gradients and partly because of the texture of the soil and the resistance to flow provided by emergent vegetation, water is stored by wetlands and released slowly over dry periods as groundwater discharge, surface flow and evapotranspiration. The storage of water on a wetland site takes the form of open water bodies and the retention of water in the unsaturated and saturated zones of the soil.

Open water

In comparison with vegetated wetlands, the resistance to flow within an open water body is very low (for a completely open water body of moderate depth it is effectively zero), and this means that open water levels throughout a wetland area are susceptible to changes brought about by manipulation, as well as by the seasonal changes that affect mires and marshes. There are a number of possibilities:

A lake or pond that discharges through a control section whose hydraulic geometry is fixed. This happens quite rarely, being confined almost exclusively to artificial reservoirs, ornamental lakes and intensively-managed wetland reserves. In this case the variation in water level is a direct consequence of climatic factors and the annual pattern of water level will consist of a seasonal fluctuation with superimposed peaks arising from rainfall and surface and short-term subsurface inputs. The installation of artificial controls is a good solution for the conservation of marginal emergent communities, the more valuable of which are sensitive to water level changes.

A lake or pond discharging through a poorly-controlled outlet, or without a point outlet at all. This is a more common situation, occurring where the control is natural or has fallen into disrepair, for instance an unmanaged outlet stream or a leaky dam, or where the water body is contained. The water level is susceptible to changes caused by weed-cutting and ditch management, by the growth and the winter die-back of vegetation and the resistance to flow at the outlet which may cause large and unpredictable rises in water level in response to rainfall events. Figure 19 shows the variation in water level in Llyn yr Wyth Eidion, the lake near the centre of Cors Erddreiniog. Llyn yr Wyth Eidion receives surface flow from a network of 19th century drainage ditches and its outfall has been manipulated, first in 1980 to improve drainage, then again in 1983 to raise overall water levels for conservation. A large area of Cors Erddreiniog drains towards the lake and water levels over most of the dipwell network give little sign of dependence on lake levels *per se*.

Ditches and natural streams. Water levels in ditches and natural streams are strongly dependent on the discharge, but the relationship between stage and discharge is not unique, except where there is a well-defined control point. Where waters are base- and nutrient-

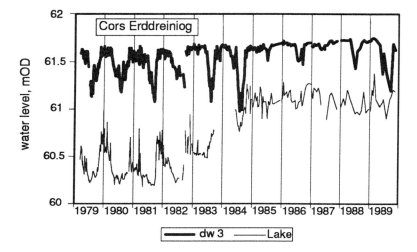

Figure 19 Variation in the water level of Llyn yr Wyth Eidion, Cors Erddreiniog, 1979-1989. Clearance of drains in 1980 lowered the winter levels in the lake. The outlet drain was dammed in summer 1983 and the control has reduced the range of variation in the years 1985 to 1989. Minor dams controlling ditch levels have raised the winter water table and reduced the summer drawdown in the peat around the dipwell dw 3.

rich, there is often a luxuriant seasonal growth of aquatics and emergents which increases the resistance to flow in the summer. Winter die-back, and the removal of the weaker-stemmed plants such as water crowfoot (*Ranunculus spp*) by spate flows, reduces the channel resistance until the following summer. The discharge and stage of the Western Cleddau at Llangloffan Fen for 1988 and 1989 are shown in Figure 20: the modest water level peak in March 1989 corresponded with a much higher discharge peak than either of those in 1988, demonstrating the effect of winter die-back on the hydraulic properties of the channel.

The water level in drainage ditches acts as a local base level for groundwater flow, but the mechanisms by which ditch water levels influence the hydrology of the wetland expanse are complex.

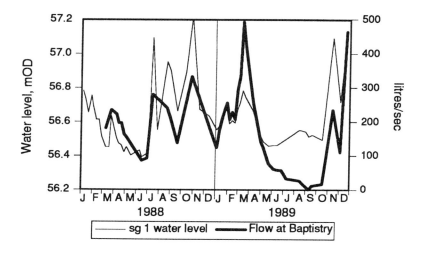

Figure 20 Estimated discharge of Western Cleddau at Llangloffan Baptistry (computed from the head over a sill), plotted with the water level measured at staff gauge sg 1, in a channel section about 100 m downstream. Six peaks of water level correspond with six peaks of discharge, but the water level to discharge relationship varies seasonally.

Soil water

Soil is a complex medium in which particles, mostly of extremely irregular form, are separated by a system of interconnected pores through which water and air can move. Water can be stored in full or partially filled pores and is held more or less tightly according to the intimacy with which it is bound to the soil particles. Of the many forces that act on water molecules in soils, gravity is the most obvious in its effects and arguably the most important: movement of water under gravity leads to the partition of the soil into saturated and unsaturated zones.

The saturated zone is that region of the soil over which the pore space is fully occupied by water. The upper part of the saturated zone is the *capillary fringe* in which the water is at less that atmospheric pressure, i.e. under tension, and is retained by the force of surface tension acting at the curved air/water interfaces in the interstices between soil particles. A dipwell inserted into the soil profile will establish an equilibrium water level at the elevation of the *water table* which is the surface within the soil at which the water pressure is equal to atmospheric pressure. Above the capillary fringe, water is retained only in the smaller pores where surface tension overcomes the gravitational force. The region of the soil where pores are occupied by both air and water is the *unsaturated zone*.

The specific yield

The saturated zone extends above the water table to a height that will depend strongly on the type of soil and its pore size distribution, finer soils having a deeper capillary fringe than coarse-textured soils. Provided that changes to the moisture content of the soil are made reasonably slowly, allowing the water to equilibrate, the removal of water from the soil will result in a decline of the water table that is proportional to the amount of water removed. The upper surface of the saturated zone will also fall by a similar amount and pores at the upper extremity of the capillary fringe will be partially dewatered. Addition of water will have the opposite effect:

$$\Delta s = \frac{S}{100} \Delta h$$

where Δs is the change in the total soil moisture storage;
S is a soil property, the specific yield, expressed as a percentage;
and Δh is the change in elevation of the water table.

Because the release of water on a falling water table is from both the capillary fringe (which is part of the saturated zone) and from the unsaturated zone, the specific yield represents an integral of soil properties between the water table and the ground surface: hence it is a function of the elevation of the water table.

Storage of water in organic soils

The water content of organic soils, which is generally high and typically can range from 80% to 95% in peat, bears little relation to the amount of water released from the soil when the water table drops. Water is stored in pores of a wide variety of sizes and in peat some of the water is contained in closed pores within partially decomposed organic matter. The specific yield of peat depends on the extent of humification and compaction. Compared with the fresh, undecomposed peats found near the surface of acid mires where the specific yield is typically above 50%, fen peats derived from reed and sedge remains, and humified peats from deep in acid mires, have higher bulk densities (based on saturated volume) and their porosity, though still very high, consists of small pores which do not drain readily (Boelter, 1964), resulting in specific yields between 10% and 20%.

The specific yield, as an integrated property of the field soil between the water table and the surface, is difficult to evaluate in the laboratory. For this reason, the water-holding capacity of soil has sometimes been measured as the difference between the water contents at saturation and at a tension of 0.1 bar, which is an approximation to field capacity beyond which the soil will not drain under gravity. This property, defined by Boelter (1969) as the *water yield coefficient*, is broadly equivalent to the specific yield but can be defined for any undisturbed soil sample. From a series of samples taken from various locations in the soil profile, it would in principle be possible to determine the specific yield for any position of the water table but the problems of sampling and the possible disturbance of the soil may introduce uncertainty. Boelter investigated the variation of the water yield coefficient of peat with other physical properties (Boelter, 1964, 1969) and found that either the bulk density or the fibre content, used as a measure of humification, could be employed as a predictor of the water yield coefficient. For fibric peats (slightly humified, 0.1 mm fibre content > 67%) the water yield coefficient was greater than 42%, for hemic or mesic peats (moderately humified, fibre content between 33% and 67%) the water yield coefficient was between 15% and 42% and for sapric peats (highly humified, fibre content < 33%) the water yield coefficient was less than 15%.

In many soils, particularly in mires where the older and more humified peat is buried and compacted by younger, fresher peat containing more open-textured organic material, the specific yield is critically dependent on the elevation of the water table, approaching 100% as the water table nears the surface. Soil moisture tensions within the unsaturated zone and the capillary fringe are rarely measurable on undrained wetland sites (Godwin & Bharucha, 1932, Gilman & Newson, 1983) except where the effects of drainage have drawn down the water table into well-compacted peat, hence the measurement of the elevation of the water table is the best index of the amount of water held in storage in the soil.

Wetland systems exhibit "feedback loops" through which hydrology and the development of the substrate can interact (Gosselink & Turner, 1978). For instance, silt trapped by vegetation builds up marsh elevations, particularly at the interface with surface streams, and reduces the incidence of flooding, in turn resulting in lessened export of organic production and the build-up of peat. Peat itself, through its excellent water-holding properties and low permeability, tends to perpetuate wetland conditions.

Layering in mires

The effect that water can have on the properties of the soil is well shown in acid mires where the peat of the mire expanse has built up beyond the influence of runoff from the catchment or the supply of water from the groundwater body in the mineral soil. An active mire system can be considered to consist of two layers (Ingram, 1978) distinguished in terms of hydrological process as the *acrotelm*, which contains the oscillating water table and possesses a high hydraulic conductivity and the *catotelm*, which has a low conductivity and a constant water content. Though seen in fens, this layer structure is best exemplified in raised mires.

The upper surface of the acrotelm can be taken as the surface of mosses in bogs and the surface of the peat in fens, but the lower surface, which separates it from the catotelm, is less clearly marked, representing the integrated effects of water table movement over a poorly-defined interval of time. In an undrained mire, only the acrotelm ever contains water in unsaturated conditions, and the presence of air in the acrotelm from time to time, and its higher hydraulic conductivity, allow more rapid breakdown of organic detritus than in the catotelm.

The lower layer, the catotelm, has been characterised by Russian authors as the "inert layer" because of the slowness of ecological processes in it, but Ingram (1982) has demonstrated that the catotelm has an important function in the water budget of the ecosystem and should not be regarded as totally inactive.

Ground movement

It is well known that the ground surface of mires is not fixed but can move in response to the amount of water stored. This phenomenon, known as *Mooratmung* ("mire breathing") was first recorded on raised bogs (see for example Ingram, 1983) but it can also be important in fens where the ground surface is supported by a considerable depth of loose peat or organic muds. Very little information has been collected on movements of the ground surface in fens: as a very rare instance, Hutchinson (1980) describes a seasonal rise and fall of the ground surface of 50 mm or more at the site of the Holme Fen Post in East Anglia. *Mooratmung* is at its most ecologically significant in floating mire or *Schwingmoor*, where the floating mat exactly matches the change in water level, but the phenomenon always acts to reduce the impact of water level variation on the wetland plant community.

Seasonal pattern of ground movement

Dipwells for the observation of water table elevation are usually inserted into the surface peat and rarely extend to a firm substrate. Measurements of depth to water table are therefore affected by ground movement if it occurs and, because the ground rises with the water table, are likely to underestimate the range of water table fluctuations. When dipwells were installed at Crymlyn Bog in 1985, it was known that the surface mat in places was underlain by very loose peat of indeterminate depth and it was decided to accompany each dipwell with a firmly-grounded datum post. Each dipwell was 2 m deep and was cased with a perforated plastic tube. The datum posts were made up of a number of 4 m sections of 19 mm electrical conduit which penetrated the peat easily and were hammered into the underlying silt and cut off just above the dipwell rim (Figure 21).

For several of the Crymlyn dipwells, regular measurements of the distance between the top of the datum post and the rim of the dipwell confirmed that the ground level varied in elevation over the year, following the water level quite closely but lagging by a few

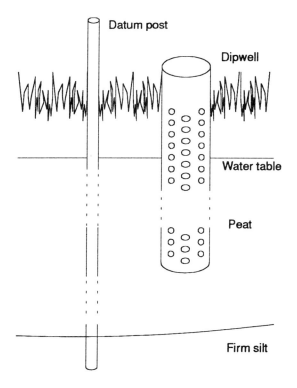

FIGURE 21   A dipwell from the Crymlyn Bog network, with its associated datum post

days and showing overall a smoother variation. Figure 10 shows the variation of water level at one of the dipwells, together with the datum-to-rim distance (note the reversed scale, which takes account of the fact that a smaller datum-to-rim distance goes with a higher ground level). Movements of the ground surface at Crymlyn, measured between 1985 and 1989, varied between 5% and 12% of the water table movement, with higher values along the axis of the site and towards the southern (downstream) end where peat thicknesses were almost certainly greatest.

The datum post system was also used at West Sedgemoor, where, despite the assurance given at one time by the River Board that peat shrinkage "has not been measurable" (Williams, 1970), the considerable depth of peat suggested that ground movement might occur, at least on a seasonal cycle. Dipwells were arranged in four straight-line transects, distributed as widely as possible across the Moor and aligned perpendicular to an important arterial drainage ditch, or rhyne. At each end of each of the transects, a specially constructed dipwell was installed (Figure 22): other wells in the transects were cased with simple perforated plastic tubes, without flanges or datum posts but otherwise similar. The casings of the end dipwells were equipped with metal and plastic flanges to engage

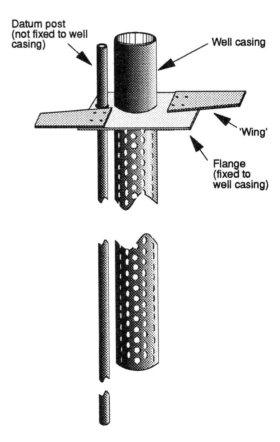

FIGURE 22 A flanged dipwell used to detect ground movement at West Sedgemoor. A 300 mm square metal plate attached to a plastic flange and buried at a depth of 300 mm below the ground surface was supplemented by metal 'wings' inserted horizontally into intact peat and bolted to the square plate. A datum post adjacent to the dipwell was intended to give a fixed height datum against which the movement of the dipwell could be measured.

securely with the upper horizons of the peat and a steel datum post was hammered through the peat to the underlying silts. In this way it was proposed to apply a correction to the water level measurements to allow for ground movement and to obtain data on the phenomenon of ground movement in its own right.

The datum posts and dipwell rims have been surveyed by levelling at six-monthly intervals by staff of NRA-Wessex between 1986 and the present, and regular readings of the vertical distance between datum posts and dipwell rims were taken between April 1988 and April 1990 at weekly or fortnightly intervals. The results of the six-monthly survey showed that there had been seasonal vertical movement of the dipwells. The datum posts were also found to have moved, with those nearest the rhynes showing a limited seasonal change in elevation which matched that of the associated dipwell, this in turn following the seasonal change in the elevation of the water table. At the other end of each transect (the "field end") there was a greater amplitude of movement of both dipwell and datum post (Figure 23). The most probable explanation for the movement of the datum posts is that they were being gripped quite firmly by soil layers near the ground surface, for instance black peat and clay layers which were found in all four dipwell transects, while sliding more easily through the permanently saturated silts at depth or flexing while gripped at top and bottom.

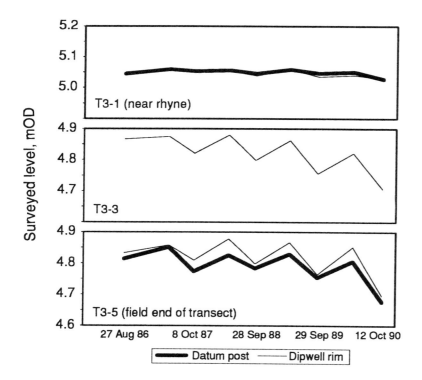

FIGURE 23 Ground movement at West Sedgemoor, detected by six-monthly levelling survey. A seasonal variation is superimposed on a general downward trend which may have resulted from a sequence of dry summers. Dipwell T3-1 was 2 m from the rhyne, dipwell T3-3 22 m, and dipwell T3-5 at a distance of 42 m.

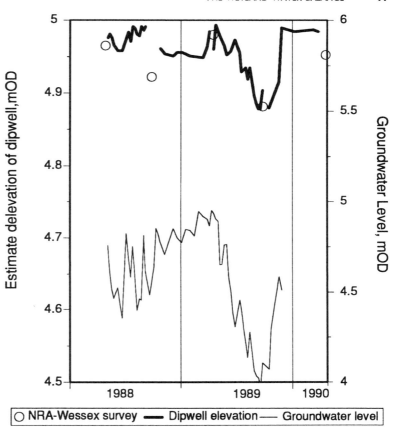

FIGURE 24 Ground level 42 m from the Middle Drain at West Sedgemoor, indicated by the elevation of the rim of dipwell T4-5, varies in response to the ground-water level. The dipwell rim elevation has been estimated from regular measurements of the vertical distance between the dipwell rim and a datum post, which itself moves with the groundwater level. The six-monthly survey results are also shown. In the figure, the ground-water level has been corrected for the vertical movement of the dipwell rim. This figure, with two similar graphs which were prepared for dipwells T1-5 and T3-5, confirmed that the spring and autumn levelling exercises coincided approximately with the maximum and minimum ground elevations reached over the year.

As the datum posts were apparently not securely fixed, it was not possible directly to determine the pattern of ground movement on a finer timescale. However, the amplitude of datum post movement at the "field end" of three transects (T1, T3 and T4) was significantly less than that of the corresponding dipwell. It proved possible for these three transects to establish a linear relationship between the six-monthly change in the difference in level between dipwell and datum post and the six-monthly change in dipwell elevation. Hence a graph of the ground level movement over the period could be derived (Figure 24).

When the estimated ground level movement was compared with the variation in groundwater levels, it was found that for dipwells T1-5, T3-5 and T4-5, the seasonal compaction was respectively 20.1%, 14.9% and 12.1% of the decline in water table, i.e. between 2.1% and 2.4% of the total depth of peat in the summer of 1989.

The causes of ground movement

When peat is drained and the water table is lowered, a fall occurs in the land surface and though this may be mitigated by careful land

and water management, it is probably true to say that a continued fall is inevitable as long as aeration prevails in the upper horizons of the peat. The phenomenon, though it was unforeseen by the great improvers of peatland, was quickly recognised as a reason for the failure to drain the East Anglian Fenland satisfactorily. Darby (1956) quotes a Colonel Dodson who addressed the Corporation of the Bedford Level in 1665:

...your Ground lying dry, the Moor Earth groweth solid to a good fruitful soyl; and it is not your Dikes bottoms which rise, but your Grounds which sink, and become much better...

There are several causes for the decline in land level:
- "Wastage" of the peat by oxidation and loss into the atmosphere as gaseous carbon dioxide and water. The entry of air into pore spaces hitherto containing water under chemically reduced conditions is responsible for this loss. The large-scale loss of about 5 m of peat from the Black Fens of East Anglia, as indicated by the progressive exposure of the Holme Fen Post (Hutchinson, 1980), can be attributed to a combination of oxidation and stripping of dry cultivated peat by wind.
- Shrinkage of the upper horizons as a result of the seasonal withdrawal of water by evaporation. The water is replaced at the end of the summer, from above by rainfall and from below by the rising water table, but it is likely that shrinkage is not completely reversible. The irreversible component of shrinkage is a process that would be expected to be most rapid immediately after drainage, slowing down as the peat in the zone of water table fluctuations approaches a stable condition.
- Loss of buoyancy of the upper horizons by removal of water. The saturated peat body is partly supported by buoyancy forces. The decline of the water table leaves a body of unsaturated peat, no longer supported by buoyancy, which exerts an increasing downward force on the peat below and hence the ground surface follows the water table downward.

Shrinkage and loss of buoyancy are both a direct result of the removal of water from the upper horizons of the peat profile and should follow a similar seasonal pattern. The difference between the two processes is shown in Figure 25. Shrinkage compresses the upper layers, which then expand as the peat is re-wetted, while loss of buoyancy compresses the permanently saturated, more fluid peat or organic mud lower in the profile.

Shrinkage of the upper horizons

Measurement of movement of the ground surface, while it demonstrates the existence of the phenomenon, provides no information about the mechanism. At West Sedgemoor, in the course of the hydrological study, it was possible to carry out a

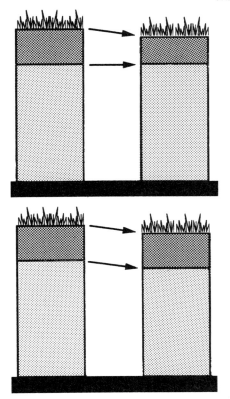

Shrinkage of upper peat on removal of water. Deeper permanently saturated layers unaffected.

Loss of water from upper horizons reduces buoyancy. Deeper more fluid peat compressed by increased weight.

FIGURE 25 The processes of ground movement - shrinkage and loss of buoyancy

simple experiment to locate the depth at which compaction was taking place. The compression of the upper horizons as a result of seasonal shrinkage can be measured by anchoring a vertical rod in the peat and measuring the displacement of the ground surface relative to the rod. Four compaction rods were installed at West

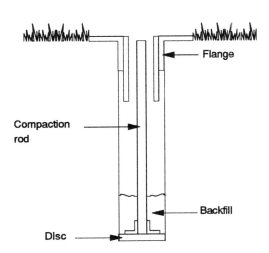

FIGURE 26 Installation of the compaction rods at transect T4, West Sedgemoor

Sedgemoor near transect T4. Plastic discs were attached to the end of lengths of metal tube, and inserted into holes drilled with a post-hole auger to depths of 0.5 m, 0.75 m, 1.0 m and 1.5 m. Each hole was partially backfilled to anchor the disc at the base of the hole and a plastic flange, with a short length of 90 mm plastic tube, was fitted at the top of the hole. The rods were free to move through the flange and the plastic tubing (Figure 26).

The vertical distance between the top of each rod and its flange was measured with a steel tape, and readings were taken at two-week intervals for the summers of 1989 and 1990. A nearby dipwell was also read on the same dates. The results are presented as Figure 27.

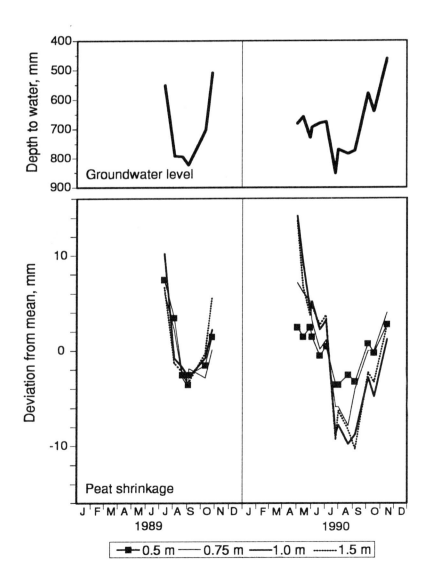

FIGURE 27 Results of compaction rod experiment, 1989 and 1990

During the summers of 1989 and 1990 there was a clear decrease in the thickness of the top 1.5 m of peat by up to 22 mm. In the 1990 summer there was a smaller decrease in the thickness of the top 0.5 m and 0.75 m but the data were too scattered to give an unambiguous relationship, for instance by linear regression. The difference between the four rods is best brought out by considering, for each rod, the variance of the results obtained over the two summers. Figure 28 shows the standard deviation of the readings for each rod, plotted against the length of the rod, and shows clearly that the majority of the shrinkage measured over the 1.5 m depth has taken place in the top metre of peat, with no shrinkage being measured in the horizon between 1.0 m and 1.5 m. This is consistent with the hypothesis that shrinkage of the upper peat occurs only in those horizons affected by loss of saturation during the summer. The figure also suggests that the amount of shrinkage of any peat horizon in the top metre is a constant proportion of about 1.3% of its thickness.

Loss of buoyancy

The ground surface movement at West Sedgemoor can be attributed to a combination of shrinkage of the surface layers, amounting to about 13 mm over the top metre during dry summers such as 1989 and 1990 and compaction of the deeper layers as a consequence of loss of buoyancy. This latter factor accounts for most of the observed ground level movement at West Sedgemoor which at sites distant from the rhynes varied between 70 and 130 mm during 1989. Data from Crymlyn Bog also supports this hypothesis: there is little drying of surface horizons there under the prevailing regime of very limited water level variation, and shrinkage of the near-surface peat would be minimal.

The identification of loss of buoyancy as the main cause of compaction helps to explain the anomalous behaviour of sites close

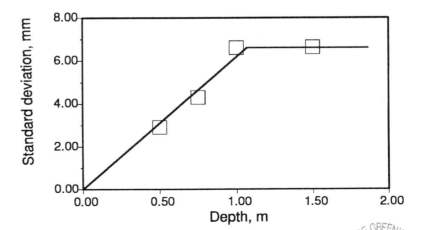

FIGURE 28   Standard deviation of shrinkage measurements, plotted against depth

to the rhynes at West Sedgemoor (see Figure 23). The seasonal range of water level in the rhynes is similar to that of the water table though the timing is very different (see for instance Figure 15), but seasonal ground movements are almost negligible near the rhynes. The buoyancy effect, due to an increase in the net weight of porous material above the water level, would not apply at the rhyne bed. Presumably there is a gradient between zero vertical movement at the rhyne bed to the full range at some distance into the field.

Inputs - rainfall

The seasonal distribution of rainfall is of some importance to the climatic template against which wetland development takes place, though in summer the rainfall often exerts less control than evaporation over the behaviour of water levels. In the west of the country, for example in Wales and the western Pennines, rainfall is greatest during the winter months and this provides the regular and reliable "topping up" that ensures that upland mires begin each year with high water levels (see Figure 7, for Cors-y-Llyn). In central and eastern England, rainfall in the summer months, largely from convective storms, can help to relieve the effects of drought on water levels but wetland water levels are restored in winter largely by runoff and groundwater flow from extensive catchment areas. The unpredictable winter rainfall, and other factors such as groundwater abstraction, introduce a random factor into winter recharge, causing inundation in some winters and an early start to the summer decline of water levels in other years. Long-term changes in the management of watercourses may also have an effect: there is a body of evidence that suggests that inundation of Wicken Fen, once common, is now quite rare owing to a lowering of penning levels of the Wicken Lode (Gowing, 1977).

Figure 29 presents simple climate diagrams for five areas of the country. The diagrams indicate the relationship between the evaporation cycle, which broadly follows air temperature, and the rainfall pattern. High rainfall with a winter peak characterises the west; high temperatures and low rainfall in the east limit wetland development to sites with large catchments. Lower air temperature in the north is an important factor leading to peat development even where the annual precipitation is low. In addition to east-west and north-south gradients, there is of course a strong dependence of climate on altitude, tending to favour peat development at high altitudes.

Interception

Of the rain and snow reaching the vegetation canopy, a proportion is intercepted by the canopy and stored temporarily on leaf and stem surfaces, where it is susceptible to the evaporation process, enhanced by increased turbulent transfer above tall crops. The

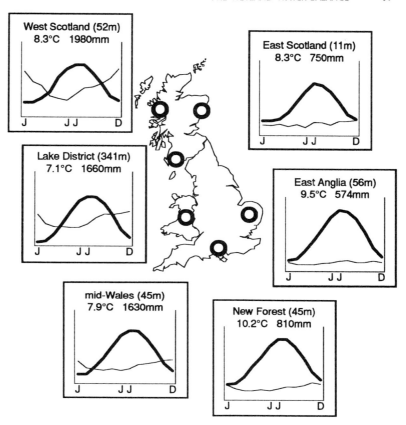

FIGURE 29 Climate diagrams (after the form of Walters & Lieth, 1960), constructed from data given by Smith (1976) for four regions of England and Wales, and by Bilham (1938) for Aberdeen Observatory and Fort William. On each diagram the monthly mean air temperature is plotted on a scale from 0°C to 20°C (heavy line) and monthly precipitation on a scale from 0 to 400 mm (light line). The annual mean air temperature and the total precipitation are also shown on each diagram.

remainder finds its way down through the vegetation, eventually reaching the ground as throughfall, which may or may not have contacted the canopy, and stemflow, which arrives at the ground as strongly localised inputs.

In coniferous forest at Plynlimon, mid-Wales, the gross interception loss (defined, following Burgy & Pomeroy, 1958, as "that portion of the rainfall that is retained on the aerial portion of the vegetation") is around 25% (Kirby, Newson & Gilman, 1991). Results from deciduous woodland are scarce but the interception loss is expected to be rather less than for evergreen trees: for example Eidmann (1959) reported interception by beech woodland averaging 8% of rainfall compared with interception of 26% by spruce in the Sauerland mountains, West Germany.

Although interception losses from forests are high enough to have serious implications for water resources, they are partially offset by the inhibition of transpiration which results from the wetness of the canopy. Only the net interception loss, after allowing for the reduced transpiration rate, is truly lost from the system. At Plynlimon, transpiration from spruce trees is between 4% and 7% of rainfall compared with between 15% and 17% of rainfall for upland grasses. Burgy & Pomeroy (1958) measured interception by rangeland grasses up to 250 mm high and found that in this case the total evaporative losses were equal for wet and dry leaf surfaces, the inhibition of transpiration equalling the gross interception loss.

On wetland sites, the losses from incoming precipitation through the interception process are difficult to quantify because of the rather variable vegetation cover and the density of near-ground vegetation. Simple net rainfall measurements on Cors Erddreiniog (Gilman & Newson, 1983) suggested that interception by tall fen vegetation could be around 50% in summer, but the results from individual sampling stations were very variable and it is obvious that rainfall quantity, duration and intensity, as well as the state of the crop, all play a part in determining the amount of interception. Although methods have been suggested which involve a comparison of the change in storage in the saturated zone with the total rainfall for each of a large number of rainstorms (Ingram, 1983), the response of the groundwater level is itself a result of several interdependent factors which give rise to considerable uncertainty and it is difficult to separate out the effect of the interception process.

The water table response

The effects of rainfall on wetland water levels are rapid and easily identifiable, the rise in water table taking place over a period that is comparable with the duration of the rainfall (Godwin, 1931; Heikurainen, 1963). Figure 30 shows the response of the water table at the lysimeter site on West Sedgemoor to an input of 46 mm of rainfall.

The net rainfall, that portion that actually reaches the ground, passes into the soil and may be incorporated into either the

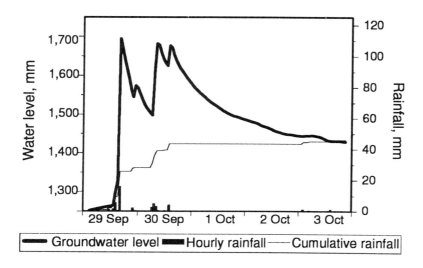

Figure 30 Hourly measurements of water table elevation at the West Sedgemoor lysimeter site show the immediate response of the groundwater body to rainfall input.

unsaturated or the saturated zone. The process of transmission of water through the unsaturated zone depends in detail on the hydraulic properties of the soil, in particular on the pore size distribution. Predictions from the conventional model for infiltration of water into homogeneous unsaturated soil, based on the Richards equation, do not always fit with observations in that transmission of water to the water table is more rapid than could be accounted for by flow through the soil matrix. Some workers suggest displacement of water already in the unsaturated zone as a mechanism for the rapid delivery of water to the water table (Horton & Hawkins, 1965). In a series of papers (Beven & Germann, 1980 & 1981, and Germann & Beven, 1981a & 1981b) a model is developed of infiltration through macropores, larger voids in the soil such as cracks and animal burrows, which penetrate deep into the unsaturated zone. In a model study, it was found that the most significant effect of macropores was in soils of moderate hydraulic conductivity where infiltration rates were greatly increased.

Intuition would suggest that, for each of a given sequence of rainfall events, the response of the water table would be related in some simple way to the rainfall (Heikurainen, 1963). Clearly, there is an underlying simple relationship but there are also interfering factors such as differential interception of large and small rainfall events, the effects of changing storage in the unsaturated zone and of errors in the measurement of both rainfall and water level change.

Using the rainfall response to determine the specific yield

Many attempts have been made to correlate rainfall events with the rise in the groundwater level, the objective in every case being the determination of the specific yield, which is the key to predicting both the natural range of water level variations and the response of the groundwater body to drainage or changes in the water balance. Godwin (1931) found that, for a sequence of eight rainstorms in October and November 1929, each centimetre of rain falling on Wicken Fen caused the water table to rise by between 7 and 12 centimetres, indicating a specific yield of around 10% in the upper 20 cm of peat.

The continuous record from Cors-y-Llyn from 1983 to 1989 was examined for rainfall peaks and a record of daily rainfalls was assembled using the data from four gauges in the Meteorological Office network. For each of 86 rainfall events, the net rainfall was estimated by subtracting a daily evaporation estimate derived from the MORECS potential evaporation, summing over more than one day for long events. The MORECS estimate (Meteorological Office Rainfall and Evaporation Calculation System) is prepared for a grid of 40 km squares on a weekly basis. Other MORECS output is mean areal rainfall, estimated actual evaporation and soil moisture deficit. Although the transpiration on a rain day could be expected to be less than potential, the interception loss from moderate-height vegetation would tend to exceed the potential rate, and it was assumed that the MORECS estimate was appropriate as an approximation to the evaporative loss for each rainfall event. The response of the water table was measured directly from the recorder chart and the specific yield for each event was evaluated as the ratio of water table rise to net rainfall, expressed as a percentage. The results for the south basin recorder are presented in Figure 31.

Considerable scatter was observed, but the estimates of the specific yield S appear to fall around a curve (the dotted line) which rises from around 20% to 80% near the ground surface, consistent with Boelter's (1969) figures for hemic (mesic) and fibric peat. A remarkable feature is the distribution of points according to the rainfall totals: small events give low estimates of S, while large events give high values. The smaller rainfall events would be affected most by inaccuracy in the assessment of evaporative losses and it is suggested that events with less than 5 mm net rainfall should not be taken into account for the estimation of S. Conversely, larger rainfall events are likely to give rise to surface runoff, which would introduce significant lateral flows into the water budget, generally tending towards an over-estimate of S and these larger events should be viewed with suspicion.

At Llangloffan Fen, a continuous water level recorder was operated on a dipwell in peat and silt about 45 m from the Western Cleddau stream. The same method was applied to the Llangloffan continuous water level record for 16 rainfall events in 1988 and 1989

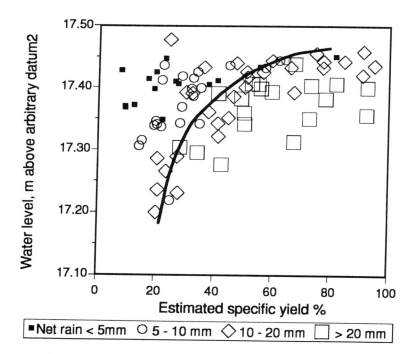

FIGURE 31 The continuous water level record for Cors-y-Llyn (south basin) contains a number of intervals during which the water level has risen in direct response to rainfall. Even after event rainfalls have been corrected by subtracting estimated evaporation figures, there is considerable scatter in the relationship, but an estimate of the specific yield can be made and it is clear that the specific yield falls off rapidly with depth.

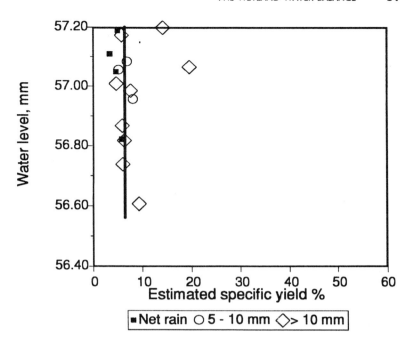

FIGURE 32 At Llangloffan Fen, the specific yield of riparian silts between 0.27 and 0.73 m below ground level could be determined to a fair degree of precision from the response to rainfall.

(Figure 32). Values of the specific yield for this material were much lower, reflecting its much higher content of fine mineral particles.

A continuous recorder mounted on a dipwell in a "litter" community at Wicken Fen provided the most consistent data set. Recorder charts between May 1984 and June 1987 yielded 59 suitable rainfall events which produced measurable peaks in the groundwater level (Figure 33). The specific yield varies between 12% (close to the Godwin (1931) estimate of 10%) at depth, to 16% near the ground surface. Gowing (1977), working from a sample of 8 rainfall events at Wicken Fen in 1976, found values of the specific yield around 15%.

The response of wetland groundwater levels to rainfall input is more clearly seen on a finer timescale: a sequence of hourly measurements of rainfall and water level, measured by the IH lysimeter on West Sedgemoor during the summer of 1990, is presented in Figure 30. The rise in water level appears to be virtually instantaneous (e.g. the very rapid rise on 29 September 1990), but the peak is followed by a swift fall. The fall late on 29 September, of about 150 mm in 6 hours, is far too fast to be accounted for by evaporative demand, which would be slack at this time of day. The highest groundwater level reached was 100 mm below ground level and surface or rapid subsurface runoff was considered very unlikely. The lysimeter itself showed similar groundwater levels and there was no runoff generated from the lysimeter. A similar decline followed the second and third peaks on 30 September and stable water levels were achieved on 3 October.

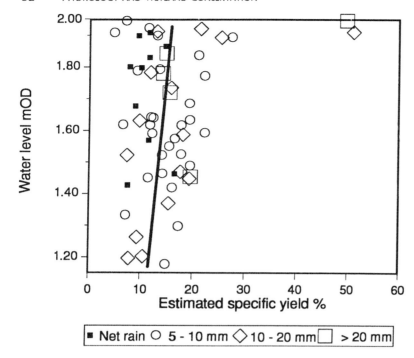

FIGURE 33 The specific yield of fen peat at Wicken Fen, determined from the response to rainfall inputs, varies between 12% and 16%.

⌐   There being no obvious mechanism for loss of water from the system, some other cause must be sought for the inconsistency between the cumulative rainfall and the change in groundwater level. It is believed that entrapment of air by rapidly infiltrating water may be the underlying cause. The groundwater level about three days after a rainfall input, by which time water and air have been re-distributed in the soil profile, is consistent with a specific yield of 25%, which agrees with estimates by other means.

Outputs - evaporation and transpiration

The literature on evapotranspiration by wetland plants, ranging from mosses to emergent plants, was examined by Crundwell (1986), revealing a continuing paucity of information which results from the difficulty in carrying out rigorous experiments in the field and from the range of wetland habitats worldwide. In comparing the results of experiments reported in the literature, Crundwell took the ratio of evapotranspiration to potential open water evaporation as a critical variable. Conflicting results have been obtained by the various studies, most of which found that evapotranspiration from a plant community could at times exceed evaporation from open water, while others had found the converse. The exact magnitude of the ratio of actual evaporation from wetlands to the potential evaporation from open water depended on the state of growth, species, climate, density and even on the methodological approach. The number of studies reported and the diversity of the plant communities studied militates against the development of a general theory to account for the results. Nevertheless Crundwell was able to formulate a number of conclusions, notably that it was possible for evapotranspiration to exceed open water evaporation because of the large area of leaf surfaces, differences in aerodynamic resistance and limited stomatal control. Most studies which had reported low values for evapotranspiration had been affected by poor experimental design, incorrect assumptions or involved plants that were not in full growth.

Notwithstanding the problem of the "oasis effect" which, by presenting a flow of dry air to the evaporating surfaces, increases the "sink strength" for evaporation, studies using isolated tanks of vegetation have been undertaken and attempts have been made to interpret the results to provide estimates of evaporative losses from infinitely extended communities. A paper by Anderson & Idso (1985) describes a study of several floating aquatics, water hyacinth (*Eichhornia crassipes*), water fern (*Azolla carolinea*) and water lilies (*Nymphaea marliac carnea*) and the emergent reedmace or cattail (*Typha latifolia*). The floating aquatics that did not rise far above the surface, water fern and water lily, gave rise to a reduction in evaporation. For water fern covering most of the surface, evaporation was 90% of the measured open water evaporation, while for water lilies the ratio decreased linearly with the area covered by pads, reaching a minimum of 85% for a (practically unobtainable) 100% cover. For water hyacinth, it was found that smaller plants (up to 0.36 m tall) growing over a large area would reduce the evaporation rate by about 10%, while taller plants lost about 1.4 times as much as open water. Results for reedmace were inconclusive but appeared to show that evaporation exceeded that from open water.

Fishponds in Eastern Europe carry marginal stands of emergent vegetation, which could be expected to transpire at a high rate in the continental summers. Smid (1975) measured evaporation rates from a dense reed stand in Czechoslovakia, using a Bowen ratio method, which relies on accurate measurements of the vertical gradients of air temperature and humidity. The evaporation rates for a vigourously growing dense reed stand were up to 6.9 mm d^{-1}, 86% higher than for open water, measured by an evaporation pan on a floating raft in a nearby pond.

Evaporation and transpiration in wetlands are not limited by the soil moisture deficit which mulches bare ground on less moist terrestrial sites with a layer of dry soil and controls the uptake of soil water by plants through an increasing soil moisture tension. During the summer there may be periods of several days without rain, during which the water table falls from day to day. Much of the decline in the water table may be attributed to transpiration demand: over the daylight hours plant roots extract water from both saturated and unsaturated zones and water is re-distributed at night to restore a quasi-equilibrium above the water table. The net effect is a decline in the water table. The maximum extent of this decline, which is an important factor determining the plant community, is itself determined by the transpiration rate and the specific yield.

Efforts to conserve wetland areas by irrigation or water level control must be backed up by estimates of the transpiration rate. Wetland plant communities comprise a range of vegetation types that have not figured largely in investigations of water use. Transpiration estimates are scarce for non-vascular plants like the *Sphagnum* mosses, for intermediate-height grassy crops such as reeds *(Phragmites australis)* and sedges *(Carex spp)* and for the open tree canopy that develops in unmanaged fenland. The key to the measurement of transpiration rates from *in situ* wetland communities lies in the physical or mathematical elimination of lateral flows which are difficult to measure and could introduce serious errors into the water balance. One solution to this problem is the lysimeter (see Chapter 4), another is to exploit the diurnal cycle of transpiration.

Diurnal fluctuations in the water table

Continuous records of groundwater level during dry periods show a diurnal fluctuation, several millimetres in amplitude (Figures 34 and 35). A relatively steep fall during the afternoon, when transpiration draws directly on soil moisture, is followed by stable water levels or a slow rise during the night. Sometimes there is a much slower fall during the night, which can perhaps be attributed to the upward re-distribution of water from the saturated zone to the unsaturated zone to replace water withdrawn by shallow-rooted plants during the day.

Though diurnal fluctuations of the water table were recorded as early as 1888 (King, 1892, quoted by White, 1932), they were not immediately attributed to evaporation. White gives the credit for demonstrating that transpiration was responsible to Prof. G.E.P. Smith of the University of Arizona, who addressed a meeting of the Geological Society of Washington in 1922.

The first detailed observations of these diurnal fluctuations in the UK were made by Godwin (1931) at Wicken Fen using a sensitive water level recorder mounted on an open pit 50 m from Drainer's Dyke, one of the principal waterways of the Fen, connected with the Wicken Lode. He described "a marked daily periodicity" which appeared between June and September; the steep daily fall between about 9 am and 6 pm was partly compensated by a rise overnight. The timing of the maximum rate of decline, between 12 noon and 3 pm GMT, shows that the lag of the decline in water table behind transpiration was small (Figure 34). Experiments with a small lysimeter containing "litter" vegetation demonstrated that the overnight rise was a result of net lateral inflow of groundwater.

Interest in the use of shallow groundwater to irrigate crops in the arid regions of the US led to an investigation of the use of the groundwater by semi-natural vegetation in the Escalante Valley,

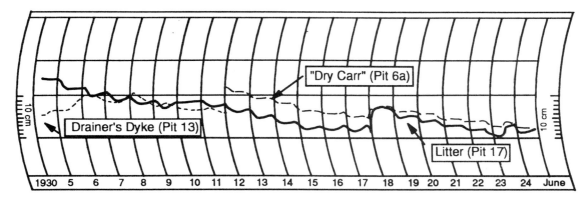

FIGURE 34 Groundwater levels recorded by Godwin (1931) at Wicken Fen in the summer of 1928 showed clear diurnal fluctuations. The records from three stations were traced on to this plot for comparison: a sequence of data from a principal dyke (Drainer's Dyke) to show that the diurnal changes were not induced by surface water and two records of groundwater levels. In "dry carr" (dotted trace in centre of diagram) there was no overnight rise while in the "litter" community (solid trace) the daily decline was partially offset by lateral flow from a dyke. Note also the steep rises in response to rainfall.

Utah (White, 1932). Diurnal fluctuations suggested themselves as a means of measuring transpiration because large phreatophytes (e.g. trees) would be difficult or slow to raise in tanks. An extensive series of measurements over a network of 75 wells, most of which were equipped with continuous water level recorders at some time during the study, was used to determine the water demand of rangeland grasses and shrubs. The depth to the water table was from 0.3 to 3.7 m and diurnal fluctuations with amplitudes between 38 and 114 mm were recorded.

The diurnal fluctuations could be clearly attributed to transpiration rather than evaporation by comparing the water table under vegetated and cleared areas. There was little diurnal change under cleared land and cutting of a crop of alfalfa reduced the amplitude significantly, this effect continuing until growth had picked up again (Figure 35). Very light rain appeared to reduce the fluctuation of the groundwater table, not by infiltration through the dry soil but by reducing the transpiration rate of deep-rooted phreatophytes. In an arid region, the overnight rise in groundwater levels, which goes some way towards replenishing the aquifer, can have its source only in the upward movement of groundwater under artesian pressure from a deeper aquifer; in the case of the Escalante Valley from sand and gravel at depth.

Provided that the specific yield is known, the amplitude of diurnal fluctuations can be used as a means of separating

FIGURE 35 Groundwater levels recorded by White (1932) beneath an alfalfa crop in the Escalante Valley, Utah. In this case the daily transpiration, which causes the steep fall during the daylight hours, is almost compensated by the overnight rise brought about by upward flow of groundwater from a deeper leaky artesian aquifer. The alfalfa was cut at the end of August (centre of diagram), and the diurnal course of water table levels, temporarily interrupted, was resumed after a few days as the crop began its regrowth.

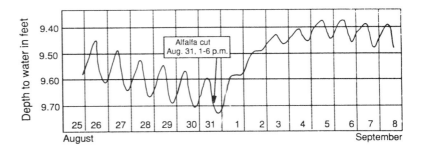

transpiration demand from net lateral inflow and hence of determining the transpiration rate. White (1932) obtained values of the specific yield from soil columns made using large cylindrical casings inserted to below water table, then sealed at the base and removed. Water was added to or removed from the sealed columns and the changes in the water table measured after 24 hours (after addition of water) or 48 hours (after removal of water). White's specific yield, which he described as "exceedingly difficult to determine", varied between 1.3% and 7.3% for seven wells and measurements of the response to rainfall for seven events appeared to agree with the 3% specific yield measured for the soil in question, a sandy clay loam. Despite being carried out under controlled conditions, the determinations of specific yield gave scattered results.

Using diurnal fluctuations to determine transpiration rates

The interest in the diurnal fluctuations lies in their relationship with both transpiration and net lateral inflow and in the information they can provide on the processes of horizontal and vertical soil water movement in wetlands. Provided that a reliable estimate of specific yield is available, both transpiration and net lateral inflow can be estimated from a continuous record of water level.

The overnight recovery of groundwater level, if present, is usually ascribed in wetlands to net lateral inflow from open water bodies or other water sources and it is assumed that the quantity of lateral flow is a function of the water table elevation and does not vary greatly over the day. This condition can be violated if the hydraulic connection between wetland and open water is too good, as for example in a flooded marsh (Dolan et al., 1984). If a mire is crossed by a dyke or other open water body where water is maintained at a high level, the lateral inflow may be expected to be proportional to the hydraulic gradient between mire and dyke and hence would remain relatively constant. At Wicken Fen, dyke levels are determined by the level of the Wicken Lode which is retained at a high level for navigation, while at West Sedgemoor the rhynes are used as wet fences in summer and are kept high by gravity drainage from the River Parrett. In both cases, the water table between the ditches assumes a bowl shape during summer and there is flow of groundwater from the ditches into the field compartments, tending to reduce the extent of the drawdown caused by transpiration.

Both Godwin (1931) and White (1932) made the assumption that the net inflow would be constant over the day and would vary little from day to day, and that the overnight rise could be extrapolated to give a 24-hour value, i.e. the rise in the water table that would have occurred in the absence of any transpiration demand. The principle is embodied in the simple equation, slightly modified from White (1932):

$$E = \frac{S}{100} \ (24r + s)$$

where E is the transpiration loss from the groundwater body over the 24-hour period from midnight to midnight;

r is the hourly rate of rise of the water table during the night: both Godwin and White favoured the use of the night before, i.e. from about 8 pm to 6 am (Godwin, 1931) or midnight to 4 am (White, 1932);

s is the net fall of the water table over the 24-hour period; and S is the specific yield expressed as a percentage.

Two continuous water level recorders were installed at Wicken Fen in May 1984 and operated until June 1987. The recorders were mounted on dipwells 19 and 20 of the Wicken network, which were in "litter" communities, distant from any dykes. Close examination of diurnal fluctuations on the records obtained from the dipwells, discernible usually between mid-June and late September, showed that the variations took three distinct forms (Figure 36). During June and the first part of July, a high evaporative demand produced fluctuations of large amplitude, characterised by a slight rise in the latter part of the morning but without the constant overnight rise which could be attributed to net lateral inflow. Godwin (1931) recorded no overnight rise in his pit 6a (the "dry carr" of figure 34) which was near to the IH recorder sites. The morning rise could be due to a re-distribution of soil water downwards from the unsaturated zone or the capillary fringe or a delayed yield after a high day-time demand from roots penetrating below the water

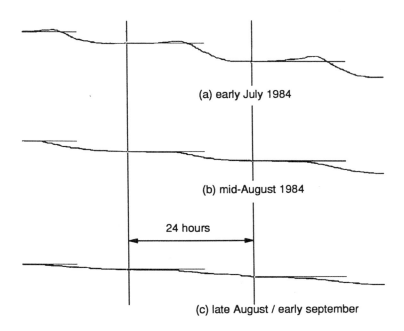

(a) early July 1984

(b) mid-August 1984

24 hours

(c) late August / early september

FIGURE 36 Three types of diurnal fluctuation recorded at Wicken Fen during the summers of 1984 to 1986. These fluctuations in groundwater level have an amplitude of several millimetres. In each case the trace extends over three days: (a) 6-8 July 1984, (b) 17-19 August 1984, (c) 31 August - 2 September 1984.

table. As the water table declined and transpiration demand was more evenly distributed between saturated and unsaturated zones, a constant night-time portion of the curve became more evident. At the lowest groundwater levels, the fluctuations became ripples on a continuous decline. This could be interpreted as a dispersion of the diurnal wave of water demand, the processes of re-distribution of soil water requiring more time to complete as the unsaturated zone became deeper. Thus the demand on the groundwater body was spread throughout the day and diurnal fluctuations became less obvious.

Of the two recorders, that on dipwell 19 was rather more reliable and the records from dipwell 19 have been used in the estimation of the transpiration demand. The results from the two dipwells, which are both distant from open water, do not show the effects of net lateral inflow. Other deviations from a constant overnight groundwater level (Figure 36) were assumed to be due to the re-distribution of water vertically within the soil profile and the transpiration demand in this case was found directly from the net daily fall in groundwater level, without the need to compensate for lateral flow. Using values of the specific yield determined from the response of the water table to rainfall, the daily transpiration rate was determined for suitable days in the summers of 1984 to 1986 (Figures 37, 38 & 39).

All three years show a transpiration maximum in June which coincides with the maximum in the potential evaporation estimate. However, early summer transpiration values were well below the potential rate. This is believed to be due to the herbaceous nature of most of the constituents of the litter community, e.g. *Molinia caerulea* (purple moor-grass) and *Phragmites australis* (common reed). In the case of reed, the annual crop does not reach maturity until late June and die-back, which takes the form of a hardening and drying of the above-ground stems and leaves, takes place in September and October (Haslam, 1970). Individual measurements of the transpiration rate of the "litter" compare well with values obtained by Godwin (1931): he measured an average rate of 3.2 mm d^{-1} in early July 1930, and a rate of 5 mm d^{-1} for the "litter" lysimeter on 9 July 1930. Other communities such as carr and young "sedge" *(Cladium mariscus)* appeared to transpire at lower rates between 0.7 and 2.0 mm d^{-1} in June and July 1930.

For a direct comparison between the estimated transpiration rates and the MORECS potential evaporation estimate, it was necessary to convert the daily rates, most of which represented dry, sunny days, to mean monthly values. Data from West Sedgemoor were used to investigate the possibility of relating the monthly mean transpiration rate to a sample of the ten highest values in each month. A daily MORECS estimate of potential evaporation for West Sedgemoor, based on a daily record of sunshine hours, had been obtained from the Meteorological Office for the summers 1987 to 1990. The average of the ten highest daily rates in each summer

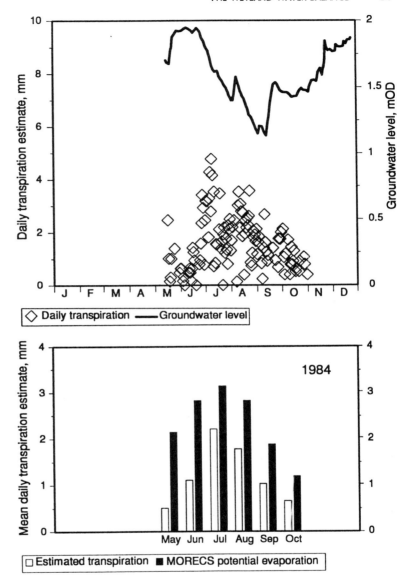

FIGURE 37 Transpiration rates for a "litter" community at Wicken Fen, estimated from the daily fall in groundwater level over the summer of 1984. Using this method, it is not possible to determine the transpiration on rain days: uncertainty in the input to the groundwater body on wet days would lead to gross errors in the output.

month (from April to October) was calculated and compared with the overall monthly average. The relationship was not linear: for months with a larger number of sunny days (i.e. midsummer months with a high average potential evaporation rate) the monthly average tended to approach more closely to the mean of the ten highest rates. The relationship between the monthly mean and the mean of the ten highest rates also showed a seasonal variation and the following empirical formula was applied to the Wicken data:

$$\overline{E}_{est} = \overline{E}_{10} (0.2542 + 0.1111 \, m - 0.0092 \, m^2)$$
$$+ \overline{E}_{10}{}^2 (0.1918 - 0.0553 \, m + 0.0043 \, m^2)$$

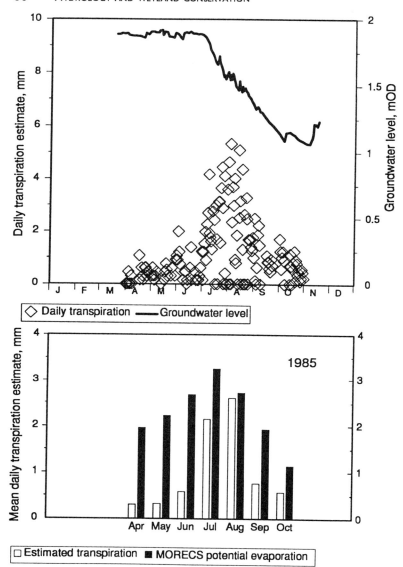

FIGURE 38 Transpiration rates for a "litter" community at Wicken Fen, estimated from the daily fall in groundwater level over the summer of 1985

where $\overline{E}_{est}$ is the estimated mean daily potential evaporation rate; $\overline{E}_{10}$ is the mean of the ten highest daily rates, and m is the month number (1=January, etc.).

The West Sedgemoor monthly values are presented in Table 2.

For the Wicken Fen data, the ten highest daily transpiration rates in each month were averaged and the formula used to determine the monthly mean. When these monthly mean transpiration rates calculated from changes in the water table level are compared with the MORECS potential estimates (Figure 40), it is clear that transpiration is very low in the early part of the summer, picks up in late June and peaks in July. The calculated transpiration values

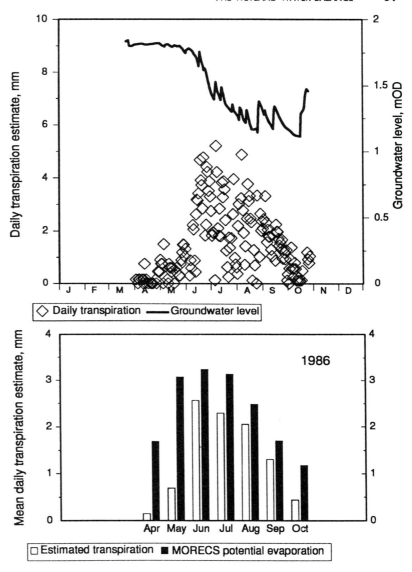

FIGURE 39 Transpiration rates for a "litter" community at Wicken Fen, estimated from the daily fall in groundwater level over the summer of 1986

are greatest as a proportion of the potential in August and fall through September and October.

The pattern of transpiration from year to year can be accounted for in terms of the pattern of management. Litter is cut on a two-year cycle and the first year's crop contains more green material than the second year's, which incorporates dead stems from the previous year. In 1985 the crop was cut in early July and growth had started again by the beginning of August. Transpiration by the green shoots growing without the shelter of a dead standing crop would be high. It is believed that the low transpiration value computed for September 1985 arose from a week-long break in the record, which

Month		1987	1988	1989	1990
		Year			
Apr	$\bar{E}_{10}$	3.43	3.21	2.80	3.42
	$\bar{E}$	2.13	1.96	1.77	2.52
	$\bar{E}_{est}$	2.35	2.17	1.85	2.34
May	$\bar{E}_{10}$	4.51	4.43	4.15	4.60
	$\bar{E}$	3.12	3.09	3.23	3.45
	$\bar{E}_{est}$	3.07	3.00	2.79	3.14
Jun	$\bar{E}_{10}$	4.12	5.90	5.78	4.95
	$\bar{E}$	2.66	3.56	3.73	2.87
	$\bar{E}_{est}$	2.67	3.97	3.88	3.26
Jul	$\bar{E}_{10}$	5.69	4.82	5.98	6.14
	$\bar{E}$	3.48	2.99	4.40	4.55
	$\bar{E}_{est}$	3.77	3.13	3.99	4.11
Aug	$\bar{E}_{10}$	4.39	4.10	4.36	5.29
	$\bar{E}$	2.91	2.50	3.38	3.29
	$\bar{E}_{est}$	2.88	2.66	2.86	3.58
Sep	$\bar{E}_{10}$	2.89	2.77	3.30	2.79
	$\bar{E}$	1.59	1.80	1.82	1.96
	$\bar{E}_{est}$	1.81	1.72	2.12	1.73
Oct	$\bar{E}_{10}$	1.28	1.31	1.94	1.58
	$\bar{E}$	0.75	0.77	1.04	0.97
	$\bar{E}_{est}$	0.68	0.69	1.11	0.87

TABLE 2 For West Sedgemoor data for the years 1987 to 1990, the mean of the ten highest daily potential evaporation rates for each month, $\bar{E}_{10}$, was compared with the monthly mean MORECS estimate, $\bar{E}$. An estimate of the mean daily rate, $\bar{E}_{est}$, using an empirical relationship derived by regression analysis, is also shown.

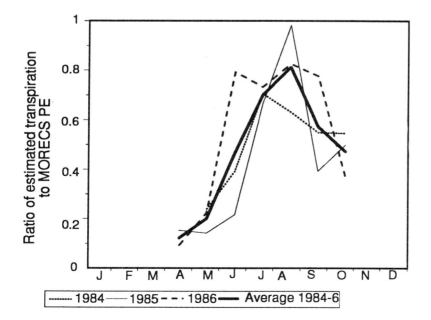

FIGURE 40 Monthly mean transpiration rates computed from changes in the groundwater level at Wicken Fen, compared with the MORECS monthly potential evaporation, for the years 1984, 1985 and 1986

coincided with high evaporation demand. The observed seasonal pattern fits well with the known course of development of the crop, with higher transpiration rates early in the year after cutting (e.g. 1984 and 1986) reflecting the smaller quantity of standing dead stems at the beginning of the year.

One of the advantages of using Wicken Fen as an experimental site is the comparative wealth of data collected over the years. The work of Godwin (1932) has been mentioned above: in the mid-1970s Gowing (1977) also installed water level recorders of a similar pattern to Godwin's and he was able to derive values of the ratio between actual evaporation and potential evaporation for the "litter" community. Gowing found that the actual evaporation, in the form of transpiration, was between 50% and 60% of the potential during the summer months in 1975 and 1976. He postulated that the ratio would rise to 80% in September and remain at 100% over the winter, as interception losses and evaporation took over from transpiration.

Throughputs

Surface flow

Even in those wetlands without standing water, so close to the ground surface is the water table that surface flow of water is a significant component of the water balance. In response to rainfall, the water table may rise until it intersects the surface and the much higher effective permeability of the litter layer and the space between plant stems allows rapid re-distribution of water across the wetland expanse. Chapman (1965) observed a clear relationship between runoff and the water table level in a mire. Once the water table dropped to 80 mm below the surface runoff became almost negligible, implying that the bulk of water movement in the peat was a fairly rapid flow through the uppermost layer, i.e. the acrotelm. Boatman *et al.* (1981) found that the "critical level", below which the decline in the water table after rainfall fell off or became step-wise, was between 24 and 76 mm below the surface in a raised mire. When this surface and near-surface flow, for instance through the undecomposed vegetation of bog-mosses or through a network of runnels between *Molinia caerulea* (purple moor-grass) tussocks, is intercepted by open drains, the still higher conveying capacity of the drains serves to remove excess water and keep the water level from rising higher. This is the secondary function of arterial drainage: to remove floodwaters as rapidly as possible and prevent the development of an anoxic zone in the upper soil.

Open drains are at their most effective when intercepting surface flow of providing an outlet for under-drainage networks. Open drains in peat are well known to have little effect in drawing down the water table in the adjoining peat (Boelter 1972). The drains investigated by Boelter were arterial drains unsupported by field drains. In the blanket peat of upland Britain, the relatively inexpensive practice of moor-gripping, intended to provide a modest improvement of rough grazing, has been shown to be similarly ineffective in lowering the water table. Nicholson *et alia* (1989) reported a drawdown of only 70 mm at the mid-point between the grips. However there is evidence to show that the most significant effects of moor-gripping are through the interception of surface and near-surface water.

Groundwater flow

The flow of groundwater through the saturated zone is governed by the hydraulic gradient and the permeability of the soil. The measurement of the permeability of peat soils has been found to be very difficult and reliance is often placed on a relationship between permeability and other properties, notably the degree of humification expressed as a point on the Von Post scale (Belding *et al.* 1975).

The permeability or hydraulic conductivity, like the specific yield, is related to the pore size distribution and in peat to the degree of humification and compaction. In spite of the strong layering of some peats, it appears that there is no consistent difference between the permeabilities in the horizontal and vertical directions. The range of values of permeability of peats is very large: Boelter (1965) found values from 0.0065 m d^{-1} for moderately decomposed fen peat to 33 m d^{-1} for undecomposed mosses. Very large permeabilities, sometimes too large to measure, are found in the upper horizons of mires where undecomposed material contains large voids. Hence much of the lateral groundwater flow must occur in these upper horizons. As an alternative to the Von Post scale of humification, the degree of decomposition may be quantified in terms of the fibre content of the peat. Boelter (1969) found that the permeability measured in the field correlated well with the content of fibres greater than 0.1 mm in length, determined by wet sieving.

A drainage ditch cut into peat provides an outlet for both surface and subsurface flow, but Boelter (1972) showed that the effect of open drains on groundwater levels in peat was limited in horizontal extent. The flow of groundwater towards a drain was highly dependent on the nature of the peat and on the layer structure of the peat. Once the water table was drawn down into moderately well-humified (hemic or mesic) peat, the low permeability meant that the zone of influence of the ditch did not extend beyond 5 m. In less humified (fibric) peat, the hydraulic gradient towards the drain extended 50 m.

Similar conclusions were reached by Burke (1961), who investigated the effects of drains on blanket peat in Glenamoy, western Ireland. In this gelatinous low permeability peat, regardless of drain spacing, the fall in groundwater level brought about by the drains was confined to a strip about 6 feet wide.

At Cors Erddreiniog, West Sedgemoor and Wicken Fen, groundwater levels along straight-line transects of dipwells have shown the seasonal pattern of variation of the hydraulic gradient. At Cors Erddreiniog the Main Drain is classified as main river and the water level is usually quite low throughout the year, depending upon maintenance of the downstream reach. At West Sedgemoor, water levels in the rhyne network are kept high during the summer and low over the winter, but rainfall events in the highland catchment fill the rhyne network and flood the Moor. The Wicken Lode has little seasonal variation, except in dry summers when the flow is insufficient to maintain the level above the sill at Upware. The obvious effects of ditches on the water table at these three sites are confined to a narrow strip and rapid changes in drain water levels have little effect unless the water rises above the ground surface. The seasonal pattern of drain levels can however induce an alternating groundwater flow, into the ditches over the winter when groundwater levels are high and outward from the ditches in the summer. The lateral groundwater flow is most important in the summer when it helps to sustain groundwater levels.

At Cors Erddreiniog, the Main Drain, having a high resistance to flow conferred by the growth of emergent vegetation, rises and falls rapidly in response to rainfall. Periodic measurements of the drain water level yielded a hydrograph similar to that of the lake, Llyn yr Wyth Eidion (Figure 19). The water level rarely approaches the level of the adjacent ground surface, so groundwater flow is always towards the drain (Figure 41).

Wicken Fen's main waterways, the Lode and connected dykes, are kept at a relatively constant level by a sluice at Upware. Groundwater levels in the adjacent Fen vary in response to the annual cycle of evaporation (Figure 42). Results from several such transects adjacent to dykes confirmed one of the conclusions of

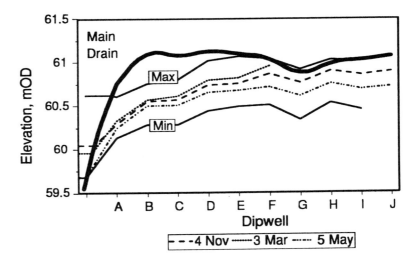

FIGURE 41 Groundwater levels during 1981 near the Main Drain at Cors Erddreiniog. Water levels were measured in a straight-line transect of dipwells (A-J) 2 m apart. Though the hydraulic gradient never changes direction, it is steeper during winter, indicating a greater flow.

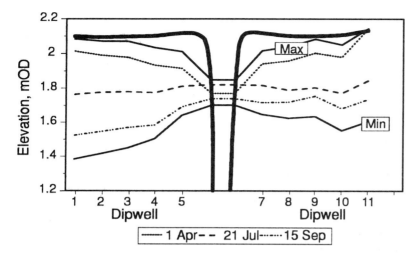

FIGURE 42 Drainer's Dyke is one of the main waterways of Wicken Fen. A transect of dipwells 10 m apart extended on both sides of the Dyke, into "litter" on the east (left of diagram) and into "sedge" (*Cladium mariscus*) on the west (right). The figure shows selected groundwater profiles in 1985, a year when both fields were cut, the "litter" in late June and the "sedge" in late August and early September. The continued fall in levels in the "litter" field may be a result of rapid re-growth over the summer.

Godwin and Bharucha (1932), that the effects of groundwater flow to dykes at Wicken were minimal beyond about 50 m.

West Sedgemoor's rhynes are strictly controlled to provide a high "summer pen" for wet fencing and a low "winter pen" for flood storage. Hence the highest levels in the rhynes, barring flood events which inundate the whole of the Moor, are found in summer, and the lowest in winter. The effect of changes in the rhyne water level and the cycle of evaporation are well shown in Figure 43, which presents selected results from a transect of dipwells in a field adjacent to a rhyne.

Such are the uncertainties in the determination of groundwater flows in wetlands, arising both from the lack of repeatability in individual measurements and from the enormous increase in

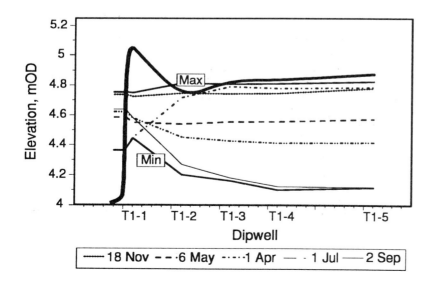

FIGURE 43 The rhynes of West Sedgemoor, with their lateral drains, form a reticular pattern which divides the Moor into rectangular fields. Results from transect T1, consisting of dipwells at 2, 12, 22, 32 and 52 m from a principal rhyne, the New Cut, show that the groundwater levels in the field are independent of the rhyne water level, except in a strip about 30 m wide. The data shown in the Figure are for the summer of 1987.

permeability close to the ground surface, that a better approach may be to regard the lateral groundwater flow as one of the unknown quantities in the water balance and to seek ways of estimating it indirectly.

At West Sedgemoor, as part of the consumptive use study undertaken for NRA-Wessex, the actual evaporation from the wet meadow community was estimated by three methods, a small lysimeter, a catchment water balance and a soil profile water balance. Over the summers of 1987, 1988 and 1989, the three methods yielded seven estimates of the actual evaporation, expressed as a percentage of the MORECS potential estimate, ranging from 77.3% to 103.4%. Ignoring two results which were regarded as outliers, the average was 95%. Using this estimate of the actual evaporation, and measured rainfall, a groundwater model was developed to simulate the movement of the water table so that the effects of changes in rhyne levels could be predicted. This model had as its parameters the hydraulic properties permeability and specific yield. These parameters were optimised to fit the simulation to measured water table levels and it was possible to use the results from the model to estimate the lateral groundwater flow.

For the purposes of the groundwater model, it is assumed that saturated flow through the peat is essentially horizontal, that it is controlled by a hydraulic gradient which is the gradient of the water table surface and that there is an underlying impermeable horizon. At West Sedgemoor the soft grey mud which underlies the peat at a depth of about 5 m is believed to be quite impermeable. The quantity of flow through a vertical panel of unit width and depth equal to the total depth of the peat below the water table is determined by the product of the hydraulic gradient and the transmissivity, which is the integral of the permeability between the impermeable base and the water table. The elevation of the water table as a function of time and space coordinates is given by the solution to the partial differential equation

$$-\nabla(T\nabla h) = -\nabla T \bullet \nabla h - T\nabla^2 h = -S\frac{\partial h}{\partial t} + q$$

which is equivalent to

$$-\frac{\partial T}{\partial h}\left(\frac{\partial h}{\partial x}\right)^2 - \frac{\partial T}{\partial h}\left(\frac{\partial h}{\partial y}\right)^2 - T\frac{\partial^2 h}{\partial x^2} - T\frac{\partial^2 h}{\partial y^2} = -S\frac{\partial h}{\partial t} + q$$

where T is the transmissivity;
$\quad h$ is the elevation of the water table, a function of t, time, and the space coordinates x and y;
$\quad S$ is the specific yield,
and $\quad q$ is the net infiltration rate (taken as the difference between

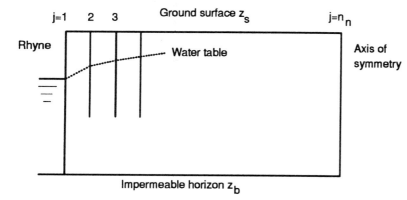

FIGURE 44 Definition sketch for the one-dimensional groundwater model. Grid points (nodes) are at 2 m intervals and in the optimisation process model predictions for nodes 2 (2 m from the rhyne), 7 (12 m from the rhyne) etc. are compared with the measured water table elevations in the corresponding dipwells.

rainfall and actual evaporation estimated from the MORECS potential).

Simplification is possible: a one-dimensional model views the dipwell transect, perpendicular to the rhyne, as part of a section through a field which has infinite extent parallel to the rhyne and is bounded by two parallel rhynes. Only half of this field need be simulated, as the centre line of the field is an axis of symmetry. The partial differential equation is solved by numerical methods which discretise the section as a grid of points at 2 m intervals (Figure 44). It was found that for most purposes a time interval of one day could be used and the model yielded daily values of the water table elevation at each of the grid points.

The variation of permeability with vertical position, and specific yield with water table elevation, could be incorporated into the digital model: for West Sedgemoor it was found necessary to have both parameters varying exponentially with depth below the surface. At West Sedgemoor, the attenuation of the effects of rhyne levels with distance was particularly rapid (see Figure 43), and it was necessary to incorporate a zone of much reduced permeability immediately adjacent to the rhyne. This reduced permeability is thought to be due to one or all of three factors: the shallow depth of the ditch compared with the full depth of the peat, sealing of the bed and banks of the rhyne or compaction of the peat near to the rhyne by heavy equipment used in the management of the ditches.

Figures 45 to 52 display a sequence of cross-sections, taken from a run of the model for transect T1 with 1987 data, and showing the cycle of groundwater levels over the year.

By early May (Figure 48) there has been a general decline in water table, although rhyne level is now high. It is worth noting that at this time of year there is a tendency for the water table in the centre of the field to lag behind the area closer to the rhyne in its decline. The model shows that a strip about 30 m wide dries out more rapidly than the rest, although very close to the rhyne water tables remain high. Sutherland & Nicolson (1986) quote a Somerset Levels farmer:

FIGURE 45 A cross-section showing the observed (circles) and predicted (dotted line) groundwater levels in Transect T1 at West Sedgemoor for 6 January 1987. Model node positions are shown by vertical continuous or dashed lines. The rhyne is at the left and the axis of symmetry of the field is the extreme right-hand dashed line (62 m from the rhyne). The upper horizontal line (just crossed by the dotted water table line) represents the ground surface at 4.8 mOD, and the lower horizontal line (immediately below the label "Section along transect") is the 3.8 mOD level.

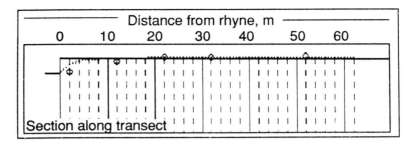

FIGURE 46 Cross-section for 3 March 1987

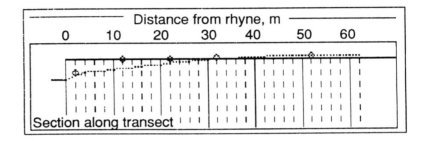

FIGURE 47 Cross-section for 31 March 1987. Although the characteristic domed shape of the winter water table persists, there has been a slight decline in overall water level, and surface flooding has all but ceased.

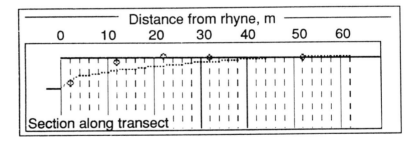

FIGURE 48 Cross-section for 5 May 1987. Transpiration in the field has caused groundwater levels to fall while the rhyne level is now retained at the summer penning level.

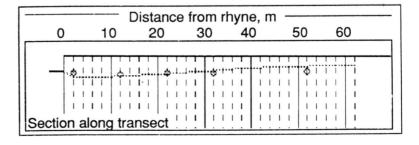

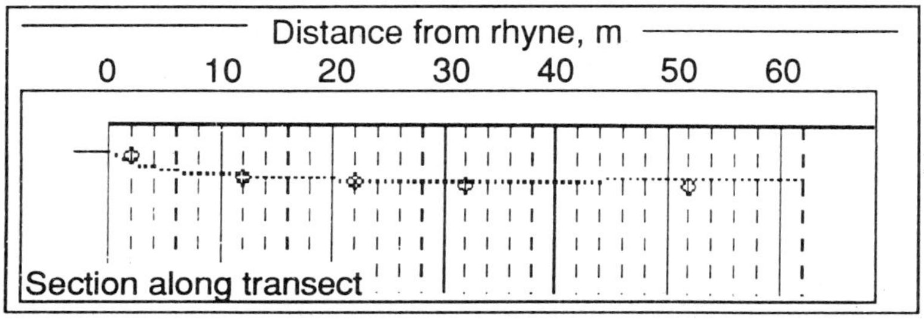

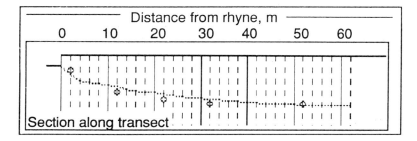

FIGURE 49 Cross-section for 2 June 1987. The decline in water table continues, but close to the rhyne the water level is maintained by flow along the hydraulic gradient which is becoming apparent in this region, and persists through the summer.

FIGURE 50 Cross-section for 1 September 1987. At this time of year the summer bowl shape of the water table is well developed. The water table in the central area of the field is almost horizontal, and the steep hydraulic gradient away from the rhyne is confined to the area immediately next to the ryne.

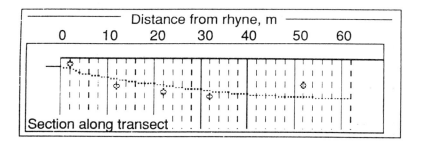

FIGURE 51 Cross-section for 6 October 1987. Water levels are beginning to rise. Uneven response to heavy rain shows in the more rapid response of dipwell T1-5.

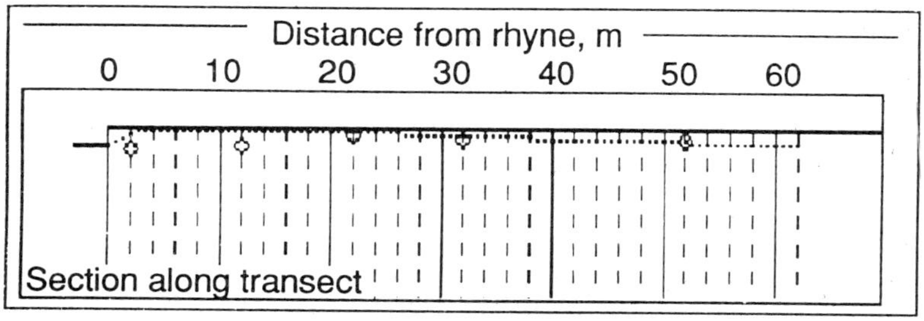

FIGURE 52 Cross-section for 1 December 1987. The winter domed shape is developing and the water table is approaching the surface. Lowering of rhyne levels at the end of the calendar year will lower groundwater levels adjacent to the rhyne, but have little effect on the centre of the field, which will soon be flooded again.

"You've got to get the water table down roughly two foot so you can actually work it. If you get less, that means the middle of the field is a day late getting dried out, after rain or whatever, than the edge. So ... you'll find you're taking smashing silage cuts off most of the field but in the middle you begin to get bogged because you haven't waited for the extra day."

The model, supported by the observed dipwell levels, suggests that the lag may in fact be much longer than one day on fields which extend far from efficiently-maintained rhynes.

Lateral groundwater flow into the peat of the "field" areas can be computed from the model output. The daily flow was calculated from the hydraulic gradient towards the ditch close to the rhyne, evaluated from the modelled water levels, and the optimised values of permeability. The variation of lateral flow from the rhyne over the year is shown in Figure 53.

While the majority of the transpiration demand by the field areas of West Sedgemoor is met by drawing on the groundwater storage, the net lateral flow from the rhynes does make a significant contribution to the water budget. When the lateral flow is summed between 17 May 1987 and 23 Nov 1987 (the period over which there is a positive lateral inflow) the total is 109 mm. This should be compared with a total rainfall of 344 mm and a total potential evaporation (from short grass) of 428 mm, over the same period.

At Cors-y-Llyn, diurnal fluctuations were used to estimate the net lateral flow as a function of depth. Diurnal fluctuations were examined and found to be mostly of types 2 and 3 (see Figure 36). The water table is generally very close to the surface at Llyn and it was concluded that upward or downward trends in the water table

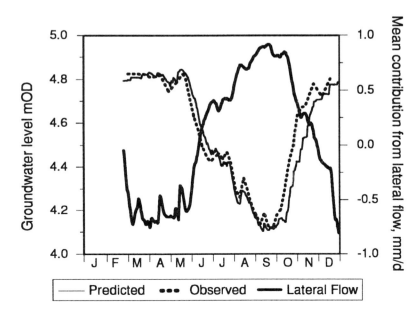

FIGURE 53 The lateral flow of groundwater from the rhyne at West Sedgemoor (transect 1 during 1987), computed from the results of the 1-dimensional model. The lateral flow component is inversely related to the groundwater level and changes sign in late spring and autumn. The contribution from lateral flow is computed as an areal average by summing the horizontal flow from rhynes on both sides of the modelled field and dividing by the area of the field. The resulting figure of just under 1 mm d^{-1} maximum should be compared with the highest potential evaporation rate of 6.7 mm d^{-1} (actual evaporation probably around 5.0 mm d^{-1}).

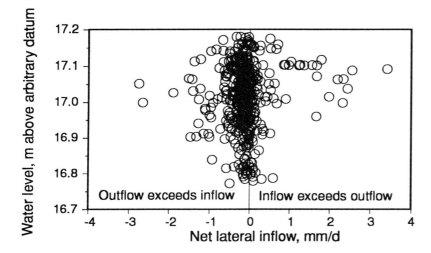

FIGURE 54 Groundwater flow at Llyn, shown by the shape of diurnal fluctuations in the water table, is usually radially outward: hence the net lateral inflow is negative. In the absence of sources of water, for instance open water bodies such as dykes, it is believed that positive values are due to local re-distribution of water.

overnight were due to lateral flow, though there were no open water bodies to account for a net lateral inflow. There is however considerable local variation in micro-topography arising from the hummock and hollow system of the mire expanse and this might explain some lateral flow in both directions. The results of extrapolating the night-time sections of more than 500 fluctuations show that lateral flow is radially outward on average: this result concurs with the oligotrophic nature of the mire (Figure 54).

CHAPTER 4

Studying the wetland water balance by lysimetry

There are many difficulties in the evaluation of water balance components on the open expanse of a wetland: the most important is the problem of assigning boundaries across which certain of the flows, e.g. groundwater flow or surface flow, will always be zero. This fundamental requirement, which underpins catchment research, can only be met on the wetland expanse by building physical barriers to prevent lateral flows. The isolation of an area of ground in this way is the first step in the creation of a *lysimeter*. A lysimeter is a soil block, with its vegetation, in which flows and storage of water can be measured while soil, vegetation and climatic conditions are kept as close as possible to the undisturbed case. Lateral flows are prevented by vertical walls but the lysimeter is usually set in surrounding soil carrying the same vegetation to avoid above-ground edge effects. By keeping the system under study *in situ*, other problems are avoided: for instance a properly designed and installed lysimeter will not suffer from the oasis effect and the vegetation canopy can develop and transpire in the same way as the surrounding undisturbed crop.

The lysimeter reduces the number of water balance components to be considered and in particular this can render the evaporation process accessible to measurement. However, if the isolated area was in receipt of a net lateral inflow, or on the other hand was generating a net lateral outflow, the presence of the lysimeter walls would change the water budget and lead to a decrease or an increase in the amount of water in store. Hence some provision must be made for maintaining ambient soil conditions by providing drainage or irrigation to keep the soil moisture equal to that of the undisturbed system.

The principal advantage of lysimeters, that they measure evaporation directly from the *in situ* vegetation, must be set against the problem of the disturbance to the soil column caused at installation. In general, only the smallest lysimeters can contain 'undisturbed' soil cores. Larger lysimeters have to be filled with loose soil and left to stabilise for a considerable period unless the walls can be sealed into a naturally occurring impermeable soil horizon. Nevertheless, despite the difficulties of installation, maintenance and interpretation, a lysimeter experiment offers a unique way of studying the behaviour of the soil-plant system on site and under naturally varying climatic conditions.

HISTORICAL BACKGROUND

While small lysimeters are appropriate for the study of vegetation on the small scale, the problem of forest water losses demands either a catchment experiment or the use of very large lysimeters. It is difficult to measure water stored in catchments and lysimeters and most estimates of evaporation are made over quite long periods of time so that storage effects cancel out. Heikurainen (1963) described a hybrid method in which changes in storage in the saturated zone were estimated from short-term fluctuations in the water table. Heikurainen constructed 20 m square natural lysimeters, using metal sheets inserted 0.5 m into humified peat to isolate stands of trees from lateral groundwater and surface water inflow. Clear diurnal fluctuations, which could be attributed unequivocally to transpiration, were observed on charts from water level recorders. The specific yield, which could be used to relate the rate of fall of the water table to the transpiration rate, was estimated from the response of the water table to rainfall after dry spells and determined more thoroughly by a series of laboratory experiments in which water was added to or removed from a column of soil contained in a watertight box. The specific yield for two sample sites was found to fall off rapidly with depth from the surface, ranging from 80% at 40 mm depth to about 14% at 220 mm depth.

Disturbance of the soil block and the time required for stabilisation following installation are important factors that have cast doubt on many lysimeter results. In wetlands it is particularly difficult to excavate to the necessary depth because of the high water table, and Bay (1966) suggested that the bottomless lysimeter, as used by Heikurainen (1963), could be employed in peat soils where the lower layers, while not completely impermeable, had such a low permeability that little water would move in or out through the open base of the lysimeter. By maintaining the head of water at approximately the same height inside and outside the lysimeter, leakage could be minimised. Bay's lysimeters were 3.05 m in diameter, and were constructed from strips of metal formed into an open cylinder 0.91 m high and driven into the peat. Water levels in the lysimeters were measured weekly, and the specific yield of the peat was determined by withdrawing metered amounts of water from the lysimeters and observing the response of the water table. Good agreement was found between lysimeter estimates of total evaporation (including interception losses) and the evaporation formulae of Thornthwaite and Hamon.

THE IH WETLAND LYSIMETER

The measurement of evaporation rates from *in situ* wetland plant communities was a central part of the IH study of hydrology and wetland conservation. The lysimeter system appeared to offer a means of estimating actual evaporation by eliminating the unknown quantity of lateral groundwater flow.

There are three main approaches to lysimeter design:

- For a small lysimeter, it is possible to cut out a soil block, with its vegetation, of a suitable size for insertion into a rectangular or cylindrical tank. The tank, provided with a drainage outlet or a weighing device, is set into the ground so that the above-ground micro-climate remains the same (Wallace *et al.*, 1982). A variant of this method is the cutting of a large core using hydraulic equipment to drive a metal cylinder into the ground: the cylinder is fitted with a base, usually by excavating alongside it, and the surrounding excavation is backfilled with loose soil.
- On a larger scale, soil can be excavated and packed into a tank. Vegetation is re-planted and the system is left to stabilise before measurements are made. The insert phytometer used by Godwin (1931) was of this type. However, the problem of re-establishment of the soil structure and vegetation makes this approach more suitable for agricultural crops than for natural communities.
- Large-scale lysimetry is possible on ground with an impermeable substrate. An area of ground is isolated from its surroundings by walls founded in the impermeable horizon. This method causes the minimum of disturbance to the growing vegetation, the area disturbed being small compared with the area of the lysimeter, and it has been adopted for forest research (Calder, 1976).

Unfortunately, in mires and marshes the installation of all but the smallest of lysimeters is complicated by several factors:

- The soil is often a deep peat, which has a low but non-zero permeability. If the lysimeter were to consist of vertical walls only, sealing of the lower edges could not be relied upon.
- The soil is saturated, with a water table very close to the surface, making excavation difficult and hindering efforts to work at the base of the lysimeter, for instance to fit and seal the bottom of a tank *in situ*. Soft ground also prevents access by vehicles and restricts the use of heavy excavating and lifting equipment.
- Peat and other chemically reduced media are vulnerable to oxidation, causing irreversible change on disturbance and desiccation. Because of macro-porosity introduced when soil is

cut and re-packed, it is not always possible to re-establish the groundwater conditions existing before the lysimeter was constructed.

● Some wetland vegetation communities, having a low nutrient status or requiring time for the establishment of root systems, are difficult to re-create. This also applies to the crop surrounding the lysimeter, which may be trampled or otherwise disturbed during installation and operation of the lysimeter.

Small lysimeters have been used successfully on acid mires but these instruments have necessarily been restricted to a scale at which the soil blocks could be cut, lifted and inserted into prepared tanks, then lowered into the excavation. Ingram (1983) contrasted the heavy engineering required to install a large lysimeter with the care that was necessary to avoid trampling damage on acid mires.

Fen and marsh vegetation is normally larger in scale and a minimum size of a square metre of ground area is required to minimise interference with roots and achieve a degree of spatial integration. The depth of the lysimeter must exceed the anticipated summer drawdown in groundwater level: a soil block of this mass (about one tonne) could not be handled, on a fen expanse, safely and without severe disturbance to both the lysimeter block and its surroundings, and the problems of sealing the base *in situ* would be formidable.

Design specification

A wetland lysimeter is a groundwater lysimeter: it must contain a soil block which is saturated for at least part of its depth. For the lysimeter to be representative of its environs, the elevation of the water table must match that in the surrounding ground, and this is usually achieved by importing and exporting water in such a way that the volumes moved can be measured. The interval at which equalisation is achieved may vary but clearly a frequent adjustment, keeping the levels close to equality at all times, is the ideal.

With frequent equalisation of water levels, there is always a very low hydraulic gradient between the lysimeter and its surroundings, and a bottomless lysimeter tank, with its base in material of low permeability, is sufficient to maintain very low leakage rates. Such a lysimeter can be inserted with the minimum of disturbance to soil and vegetation, requiring only the insertion of a set of vertical walls into pre-cut slits in the soil. As the walls can be assembled on site, the horizontal scale is arbitrary to an extent, depending on cost and on the dynamics of the water level control system.

A system for automatic equalisation of water levels must comprise the following elements:

● Water level measurement at two points, to represent conditions inside and outside the lysimeter;

- A means of transferring water into and out of the lysimeter, the direction and the amount of the transfer to be controlled by the difference in water levels;
- Precise metering of the amount of water transferred, so that it can be taken into account in the interpretation of the results.

As the control of water levels requires precise measurement, and the response in the form of pumping of water is best controlled by electrical means, the IH wetland lysimeter was designed around a microcomputer which could be programmed to carry out all the operations above and to store the data on magnetic tape for later interpretation by computer.

Operation of the IH instrument

The prototype instrument was in operation on Skew Bridge Bog, a soligenous mire at Llangurig, Powys, between spring 1986 and summer 1989. The walls, of PVC sheet enclosing a square of land of area 1.43 m², were installed in the spring of 1985 and the installation and trials of the control system took place in that autumn. The walls of the lysimeter extended to a depth of 0.75 m and were joined at the corners by alloy angle and sealed with a silicone sealant. The walls were assembled on site into a bottomless box. Their lower edges were chamfered so that the assembled box could be inserted by hammering into a pre-cut slot in the peat (Figure 55).

Water table elevations were measured in the centre of the block and at a point outside the walls by dipwells fitted with potentiometric float-operated water level sensors. The resolution of the water level measurements was about 0.2 mm. Hourly water level measurements were stored on magnetic tape. Additionally, 15-minute measurements were used as inputs to the control system, which operated

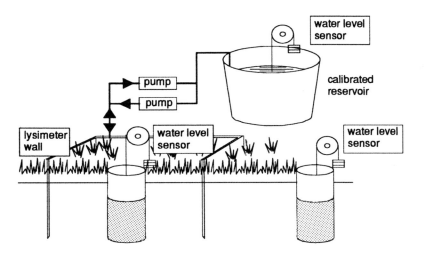

FIGURE 55 Layout of the IH wetland lysimeter

two battery-driven peristaltic pumps to move water into and out of the lysimeter. Depending on the difference in water levels, the appropriate pump was operated for an interval of one minute. Using this system the difference in water levels between the lysimeter and its surroundings could be maintained to better than 0.5 mm.

It was necessary to measure the quantity of water transferred. This was achieved by pumping into and out of a calibrated reservoir tank whose level was monitored by a third water level sensor. Control of the pumps was effected by an Epson HX-20 micro-computer which also served as a data logger, recording hourly values of water levels in the lysimeter, outside it and in the reservoir. The program for data logging and control was written in BASIC.

A similar lysimeter system was installed at West Sedgemoor in the summer of 1989 and operated for a trial period during September and October 1989. Instrumental problems were encountered in early 1990 and it was late June 1990 before the equipment was functioning correctly. The instrument continued to operate until November 1990.

The West Sedgemoor lysimeter was located near dipwell transect T4, a short distance from a lateral drain, and the effect of this drain in providing a net lateral inflow to the peat around the lysimeter could be detected in the records.

Determination of specific yield

There are various ways of interpreting the data obtained from the lysimeter and the use of a microcomputer allowed an unprecedented degree of flexibility in the sequence of operations and the choice of parameters. Though the underlying function of the pump control system was to maintain equality of the water levels, the lysimeter also offered an automatic method, that did not rely on manual intervention or rainfall events, for determining the specific yield of the soil.

A knowledge of the specific yield is essential to the calculation of evaporation using either the lysimeter data or any other information on diurnal fluctuations in water level. Using the control system built into the lysimeter, it was possible to determine the specific yield by pumping a known quantity of water out of the enclosure and observing the change in water level within the lysimeter. Allowance for leakage that occurred under the walls of the lysimeter while there was a difference in water level could be made using a method which was originally developed for the computation of leakage of heat during specific heat determinations (the cooling correction).

In principle, the pumping test could be applied each night to give a range of measurements of specific yield at various depths. However, there are some signs that the pumping test can interfere with the daylight measurements, particularly in late summer when the water table is low, and the frequency of pumping tests was reduced to one night in five during 1989.

The manipulation of water level for the determination of specific yield was carried out under automatic control during the evening and night, when any change in groundwater level caused by evaporation had ceased and levels were relatively constant. Water was pumped from the lysimeter for one minute in each quarter-hour between 2000 and midnight GMT, and the fall in water level in the lysimeter was recorded. Between midnight and 0400 the pumps were switched off, and leakage caused a (generally slight) rise in level within the lysimeter. From 0400 onwards, water was pumped back into the lysimeter until levels inside and outside were

TABLE 3 Results from the Skew Bridge Bog lysimeter for 5 May 1987. All water level readings are expressed in mm above arbitrary datum.

Date	Time	Water levels		
		Lysimeter	Mire	Reservoir
4 May 1987	2000	1691.0	1691.6	420.3
	2100	1685.8	1691.6	430.1
	2200	1682.0	1692.2	439.7
	2300	1678.6	1692.4	449.1
	2400	1676.0	1692.4	458.7
5 May 1987	0100	1677.8	1692.4	458.7
	0200	1679.6	1692.8	458.7
	0300	1681.0	1692.8	458.7
	0400	1682.0	1692.8	458.7
	0500	1682.8	1692.8	458.7
	0600	1688.6	1692.8	450.3
	0700	1692.4	1692.8	443.3
	0800	1693.0	1692.8	440.9
	0900	1692.6	1692.8	440.9
	1000	1691.0	1691.4	440.5
	1100	1689.8	1690.4	440.7
	1200	1688.4	1688.8	440.3
	1300	1688.2	1687.2	438.2
	1400	1686.6	1685.8	438.2
	1500	1684.1	1684.7	440.6
	1600	1683.3	1683.7	440.6
	1700	1682.3	1683.1	439.4
	1800	1682.7	1683.3	439.4
	1900	1682.6	1683.4	439.4
	2000	1682.8	1683.4	439.3
	2100	1677.8	1683.6	448.9
	2200	1674.2	1684.2	457.8
	2300	1671.3	1684.8	467.5
	2400	1668.9	1684.8	476.4
6 May 1987	0100	1670.1	1685.2	476.5
	0200	1672.1	1685.4	476.7
	0300	1673.1	1685.4	476.7
	0400	1673.9	1685.4	476.7
	0500	1674.7	1685.4	476.5
	0600	1681.5	1685.8	468.9
	0700	1686.6	1686.4	460.1
	0800	1686.6	1687.2	459.9

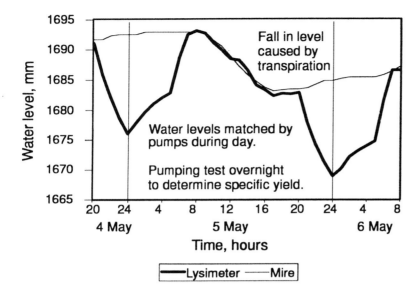

FIGURE 56 Data from the Skew Bridge Bog (Llangurig) lysimeter for 5 May 1987. Water levels from the lysimeter and the mire outside the lysimeter enclosure are shown. Note the drawdown of lysimeter water level by pumping between 2000 and 2400, the partial recovery as water leaks into the lysimeter between 2400 and 0500, then the steep rise as reservoir water is pumped into the lysimeter to equalise water levels.

again equal. By 0700 the levels had usually returned to equality.

The water level readings were checked periodically using an electric contact gauge and all readings were relative to an arbitrary but fixed datum. When necessary, for example after an instrument breakdown or battery failure, water levels inside and outside the lysimeter were equalised using a siphon.

A sample day's recordings from the Llangurig lysimeter is shown in Table 3. The data are also plotted in Figure 56. For each day the calculations are performed on batches of data extending over 36 hours (37 readings).

For the determination of specific yield, it is required to find the change that would have occurred in the lysimeter groundwater level in the absence of leakage under the walls. The key to this is the behaviour of the lysimeter water level over the interval when pumping has stopped, and there is no input to the lysimeter except leakage. As groundwater flow is described by Darcy's equation, which is linear, it may be assumed that the rate of leakage is proportional to the difference in levels:

$$q_{leak} = \alpha(h_{mire} - h_{lys})$$

where q_{leak} is the rate of leakage; h_{mire} is the groundwater level outside the lysimeter; h_{lys} is the groundwater level inside the lysimeter, and α is a constant of proportionality.

It follows that the cumulative leakage over any given interval (t_1, t_2), the integral of the rate of leakage with respect to time, is proportional to the integral of the difference in levels with respect to time:

$$Leakage\Big|_{t_1}^{t_2} = \int_{t_1}^{t_2} q_{leak}dt = \alpha \int_{t_1}^{t_2} (h_{mire} - h_{lys})dt$$

The difference in levels, corrected if necessary for the slight imbalance at 2000, is calculated for each hour from 2000 to 0700 and the integral is approximated using Simpson's rule for the intervals between 2000 and 2400 and 2400 and 0400 and the trapezium rule for the interval between 0400 and 0700. These integrals are denoted by A_1, A_2 and A_3 (see Table 4).

Leakage into the lysimeter over the first time interval, between 2000 and 2400, causes a reduction in the drawdown by an amount Δh. This leakage is proportional to Δh:

$$Leakage\Big|_{2000}^{2400} = \int_{2000}^{2400} q_{leak}dt = \alpha A_1 = \frac{S}{100}\Delta h$$

The leakage into the lysimeter over the "no pumping" time interval, from 2400 to 0400, is proportional to the rise in the lysimeter level, the constant of proportionality being the specific yield:

$$Leakage\Big|_{2400}^{0400} = \int_{2400}^{0400} q_{leak}dt = \frac{S}{100}(h_{lys}(0400) - h_{lys}(2400)) = \alpha A_2$$

Eliminating a and S,

$$\Delta h = \frac{A_1}{A_2}"h_{lys}(0400) - h_{lys}(2400)"$$

Thus, in the absence of leakage, the fall in the lysimeter groundwater level between 2000 and 2400 would have been

Time	Drawdown, mm	Integral
20:00	0.0	-----
21:00	5.2	
22:00	9.6	A_1 = 36.2362
23:00	13.2	
24:00	15.8	-----
1:00	14.0	
2:00	12.6	A_2 = 50.7507
3:00	11.2	
4:00	10.2	-----
5:00	9.4	
6:00	3.6	A_3 = 11.6116
7:00	0.0	-----

TABLE 4 Differences in water level used in the process of correction for leakage

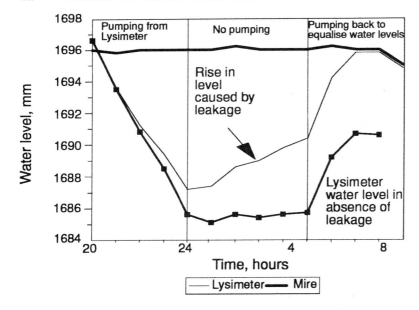

FIGURE **57** When water levels are not equalised, there is a leakage under the lysimeter walls, the magnitude of which is proportional to the difference in levels. The leakage may be estimated by a simple numerical method. The success of this method may be judged by plotting the predicted water level in the lysimeter if no leakage had taken place.

$$(1691.0 - 1676.0) + \frac{36.2}{50.7} \times (1682.0 - 1676.0) = 19.3 \text{ mm}$$

Figure 57 shows the leakage correction applied to the lysimeter water level over the interval 2000 to 0800 for a sample day.

The water is pumped out of the lysimeter into the calibrated reservoir. The rise in reservoir level over the period 2000 to 2400 is

$$458.7 - 420.3 = 38.4 \text{ mm.}$$

The quantity of water represented by this rise is equal to the product of the rise and the surface area of the reservoir. Because the reservoir is not quite cylindrical, the surface area is a function of water level, and this function was established by a number of calibration exercises, in which a known quantity of water was added to the reservoir, and the change in level observed. The relationship obtained for the Llangurig lysimeter was

$$\text{Surface area} = 0.1187 + 0.000502X + 0.000627X^2$$

and for the West Sedgemoor lysimeter

$$\text{Surface area} = 0.1205 + 0.000469X + 0.000603X^2$$

where X = water level/100.

The surface area given by this formula is in m². Figure 58 shows the calibration curve for the Llangurig lysimeter.

Returning to the example, over the interval 2000 to 2400, the

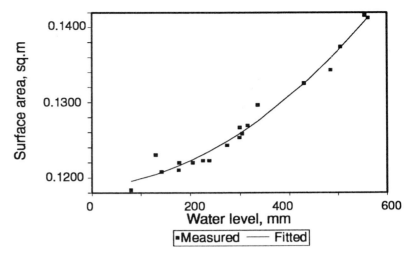

Figure 58 Reservoir calibration curve for the Skew Bridge Bog lysimeter

mean reservoir level is 448.1mm and the surface area for this level is 0.1336 m². The rise in reservoir level is therefore equivalent to the addition of a quantity

$$38.4 \times 0.1336 = 5.130 \, l.$$

The area of the lysimeter is 1.43 m²: hence the pumping has caused the dewatering of a volume of soil of

$$19.3 \times 1.43 = 27.60 \, l.$$

The specific capacity, expressed as a percentage, is

$$\frac{5.130}{27.60} \times 100 = 18.6\%$$

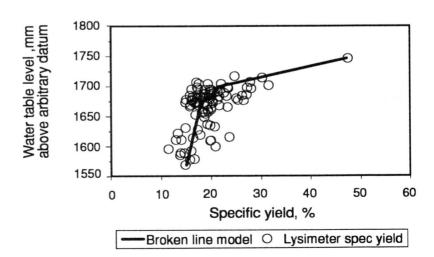

Figure 59 Specific yield of peat at the Skew Bridge Bog lysimeter, from measurements taken by the mire lysimeter in 1986 to 1989. The broken line expresses the variation of specific yield with water table elevation.

Over the summers of 1986, 1987 and 1988, a total of 96 determinations of specific yield were performed at Skew Bridge Bog and the results are plotted in Figure 59.

The dependence of specific yield on the elevation of the water table is clear. Unfortunately, owing to the infrequency of very high water table levels, there are few measurements close to the ground surface where specific yield is at its highest. However, the method ensures that a large number of determinations are carried out over the modal range of elevations.

The relationship is well expressed by the broken line:

$$S = 15 + 3 \; \frac{h - 1575}{1670 - 1565} \qquad \text{for} \qquad h \leq 1670$$

$$S = 18 + 4 \; \frac{h - 1670}{1700 - 1670} \qquad \text{for} \qquad 1670 < h \leq 1670$$

$$S = 22 + 25 \frac{h - 1700}{1745 - 1700} \qquad \text{for} \qquad 1700 < h$$

and this relationship was used in the computation of transpiration from the lysimeter.

At West Sedgemoor, because of the shorter period of observations and the restriction of pumping tests to one day in five, there were fewer determinations of the specific yield. Thirteen measurements were taken, ranging from 12.5% to 33.9% and these are plotted in Figure 60. Again a broken line relationship would appear to be suitable, in view of the limited range of variation of specific yield to be expected in fen peat. The results are in good agreement with optimised values of the specific yield and its variation with depth, derived during the modelling study described in Chapter 3.

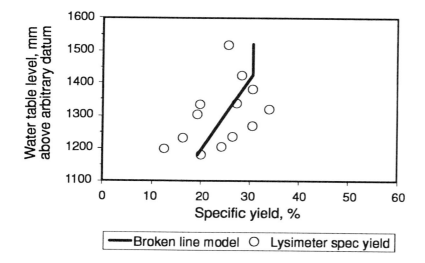

FIGURE 60 Specific yield of peat at the West Sedgemoor lysimeter, from measurements taken by the mire lysimeter in 1989 and 1990. The broken line expresses the variation of specific yield with water table elevation.

The broken line relationship between specific yield and water table elevation

$$S = 47.6 - 0.04562(1800 - h) \qquad \text{for} \quad h \leq 1425$$
$$S = 30.49 \qquad\qquad\qquad\qquad \text{for} \quad 1425 < h$$

was used in the computation of transpiration rates from the West Sedgemoor lysimeter.

Analysis of results to determine transpiration rates

Water lost from a wetland surface by transpiration causes a decline in the water table over the summer, but the picture is complicated by lateral groundwater and surface water flows which tend to compensate in part for evaporative demands. Water levels typically fall during the day but recover partially during the night. It is thus incorrect to attempt to calculate evaporation from the daily decline in water level without taking account of the net lateral flow.

Raised mires are domed in shape because of a continuous flow of groundwater radially outwards; fen water tables tend to alternate seasonally between dome-shaped and saucer-shaped. The curvature of the water table indicates a 'divergence term' in the water balance; a dome-shaped water table has a net lateral outflow while areas with a saucer-shaped water table have a net lateral inflow. The IH wetland lysimeter performs separate measurements of evaporation and lateral inflow or outflow.

Consider a lysimeter operating over a day of zero rainfall and moderately high evaporative demand. Without the importation of water via the pumping system, the lysimeter water level would drop further than that outside which was maintained in part by lateral inflow. The quantity of lateral inflow can be determined by continuous equalisation of the water levels and measurement of the pumped volume, the latter being achieved by monitoring the level in the reservoir. The fall in water level in the lysimeter, equal to that outside, when multiplied by the specific yield of the soil, gives a number equal to the evaporation minus the lateral inflow. Thus if the specific yield is known, both transpiration and lateral inflow can be determined. The use of this method, the *direct method*, is described in detail below.

A simpler *indirect method* is available, particularly if the diurnal fluctuations of water level are strongly defined. This method is also applicable to results from a single water level recorder on a dipwell, provided that the specific yield can be estimated by some means (see Chapter 3). Lateral inflow is assumed to be essentially constant over the 24-hour period and an estimate of the lateral inflow is made from the product of the overnight rate of rise and the specific yield. The transpiration loss is the sum of this lateral inflow and the quantity obtained by multiplying the daily fall in water level by the specific yield.

Direct method

Though the direct method makes full use of the potential of the lysimeter to measure the flows of water necessary to maintain equality of levels, and hence offers a direct measurement of the net lateral inflow, results obtained in this way are rather suspect when taken on days when there has been a pumping test to determine specific yield. The reason for this is that the leakage induced by the pumping test is comparable in magnitude with the net lateral inflow. Though the difference between transpiration and net lateral inflow, which shows itself as a change in the water table, is measurable with relative precision, the uncertainty in the leakage leads to a wide margin of error in the net lateral inflow. However, it is instructive to compute the net lateral inflow and transpiration using both methods as there are occasions when one or other of the methods, but not both, may be invalidated by rainfall.

It is impossible to derive results from the lysimeter on rain days, or more strictly when there is rainfall during the hours of heavy transpiration demand, owing to the uncertainty about interception and storage in the upper soil and the likelihood that rainfall will exceed transpiration several times over. On a dry day the three terms in the water balance of the lysimeter are transpiration, leakage (induced by the pumping test if any during the night) and pumping.

The total leakage induced by the difference in levels during the pumping test, using the notation of the previous section, is

$$\alpha(A_1 + A_2 + A_3) = \frac{S}{100 \times A_2}\{h_{lys}(0400) - h_{lys}(2400)\}(A_1 + A_2 + A_3).$$

For the example day, 5 May 1987, this amounted to 2.33 mm. The mean level in the reservoir between 2000 on 4 May and 2000 on 5 May was 444.1 mm, and the fall in reservoir level over the same period was

$$-(493.3 - 420.3) = -19.0 \text{ mm}.$$

The surface area of reservoir at the mean water level was 0.1333 m², so the volume of water pumped out over the 24-hour period was

$$-19.0 \times 0.1333 = -2.533 \, l$$

or

$$-\frac{2.533}{1.43} = -1.77 \text{ mm}$$

over the area of the lysimeter. The net lateral inflow, excluded by the lysimeter walls, would have been equal to the sum of the leakage during the pumping test and the amount pumped from the reservoir, i.e.

$$2.33 - 1.77 = 0.56 \text{ mm}.$$

The transpiration is equal to the decrease in groundwater stored in the mire, less the net lateral inflow:

$$-(1683.4 - 1691.6) \times \frac{20.0}{100} - 0.56 = 2.20 \text{ mm}$$

It is clear that some uncertainty will attach to the net lateral inflow, which on this particular day was the difference between two similar quantities, and in extreme cases this may invalidate the method.

Indirect method

The indirect method does not use water level data from the lysimeter or the reservoir, but relies on a good estimate of the specific yield, obtained by a series of pumping tests. For this reason it provides a more robust estimate of both the net lateral inflow and the transpiration rate, although it is still subject to error on days when the water level is raised by rainfall. Usually, the effects of rainfall are obvious if water levels are examined in graphical form and days with visible rainfall inputs should be discarded.

Given a day with a clear diurnal fluctuation (5 May 1987 again serves as a good example) it is possible to estimate the daily change in water levels that would result from lateral inflow and to take this into account when estimating the transpiration.

The rate of rise in water level over the hours of darkness is calculated by an approximate linear regression method, based on the division of the data into four groups, comprising the measurements between 2000 and 0200, 0300 and 0800, 2000 and 0200, 0300 and 0800. For the sample day, 5 May 1987, the average rate of rise was 0.1575 mm h^{-1}, equivalent to 3.78 mm d^{-1}.

The specific yield is calculated from the mean lysimeter water level using the broken-line relationship; a figure of 20% is obtained.

The fall in level over the daylight hours that would have been caused by transpiration in the absence of lateral inflow is computed from the change in water level over the 24-hour period, i.e. between 2000 and 2000, and the 24-hour rise caused by net lateral inflow. The transpiration is calculated from the equation:

$$E = \frac{S}{100}(24r + s)$$

$$E = \frac{20.0}{100} \times (24 \times 0.1575 + 8.2) = 2.40 \text{ mm}$$

Comparison of results with potential evaporation

It was possible to compute estimates of the net lateral inflow and the daily transpiration rate for 92 days during 1986, 1987, 1988 and 1989 from the results of the Skew Bridge Bog lysimeter and these figures have been compared with the Penman estimate of evapotranspiration for short grass, computed from automatic weather station data. An automatic weather station was operated at Skew Bridge Bog over the summers of 1986 and 1987 and fill-in data for gaps in this record and for the summers of 1988 and 1989 were obtained from long-term automatic weather stations operated by IH Plynlimon at Cefn Brwyn (SN 826838) and Eisteddfa Gurig (SN 803853), by setting up a linear relationship between these stations and that at Skew Bridge Bog.

The West Sedgemoor lysimeter provided a total of 74 evaluations of daily transpiration in 1989 and 1990. For the purpose of comparison, estimates of daily potential evapotranspiration were obtained from the Meteorological Office for West Sedgemoor.

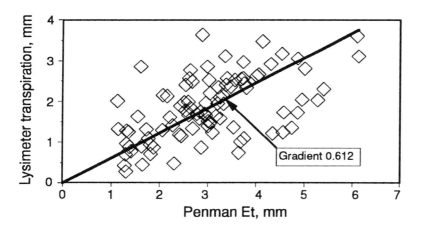

FIGURE 61 Comparison between daily transpiration measured by the Llangurig lysimeter (1986-1989) and potential evapotranspiration derived from automatic weather station data

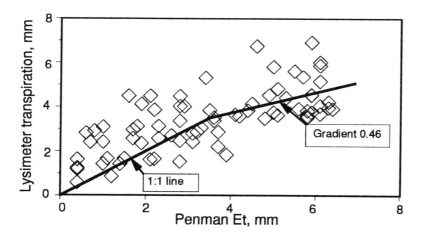

FIGURE 62 Comparison between daily transpiration measured by the West Sedgemoor lysimeter (1989-1990) and daily potential evapotranspiration derived from the monthly MORECS estimate using sunshine records

These had been derived from the monthly MORECS potential evaporation figures by applying a daily factor depending upon sunshine hours recorded at Yeovilton.

At Llangurig, there appeared to be a linear relationship between the lysimeter transpiration, in the months of April to October, and the potential evapotranspiration on a daily basis: values were scattered around a ratio of 0.612 (Figure 61). No systematic variation in this ratio over the summer was detectable in the data.

Comparison of the lysimeter estimates of actual transpiration at West Sedgemoor with the potential evapotranspiration derived from the MORECS figures showed a similar scatter of points but higher values of potential evapotranspiration were not matched by high actual transpiration as measured by the lysimeter. Though transpiration apparently equalled or exceeded the potential for low daily rates, for rates above about 3.5 mm d^{-1} there was a deviation from the 1:1 line, and points above this level were better represented by a line of gradient 0.46. Taking the daily MORECS figures for the summer of 1990, this would mean that the actual transpiration, assuming no interception losses, would total 97% of the potential. This figure compares well with the average figure of 95% which was one of the main conclusions of the study for NRA-Wessex.

CHAPTER 5

Summary and conclusions

In recent years there has been an upsurge of interest in the conservation of wetlands as it has become clear not only that wetlands in the UK are of great habitat value, often on an international scale, but that they have been reduced in number and extent by development pressures and that even the small remnant are threatened from many directions. Wetlands cannot survive if the hydrological template within which they have developed over the millennia since the last glaciation is disrupted by diversion of water supplies or drainage, but the protection of whole catchments for the sake of wetland reserves is an unrealistic objective.

The pressing need for protection of the remaining wetlands has evoked the commitment of large funds to land purchase by conservation bodies, and more determined opposition to the development or improvement of wetlands themselves and of adjoining land. Successful opposition to development, leading for instance to the calling-in of a planning application by the Secretary of State, can purchase a brief respite, and the designation of land as SSSI can provide a degree of protection against changes in land use practices. The probability is that compromise solutions will become more and more common in the future, with developers committed to measures for conservation or the provision of land for habitat creation as compensation for the loss of wetland sites. Wetland conservation must take more account of the hydrological background to each site if it is to progress beyond the sometimes over-optimistic designs on the one hand of conservationists determined to fence out the threat and on the other hand of developers anxious to appear 'green'. Wetland sites are there for historical reasons, one or more hydrological processes having conspired to provide the excess water that is a feature of wetland habitats. Without a clear understanding of the hydrology of the site it is impossible to predict the effects of changes in adjoining land, to confirm the long-term viability of existing or created wetlands, or to improve on the management of the land to increase its habitat value.

This volume has attempted to explore the principles of hydrology and wetland conservation and to point the way for future use of hydrological methods in the management and interpretation of wetland sites. A small number of sites were selected to demonstrate the similarities and differences between the various wetland types represented in the UK, and above all to show the value of long-term monitoring of hydrological variables as part of the management of wetland sites. Once the hydrology of the site is understood, it is possible to make informed decisions about the likely response of the site to outside pressures, ranging from peripheral drainage to climate change, and about the management of the site to counteract threats and maintain conservation value.

FRESHWATER WETLANDS IN THE UK

Wetland sites, though they have a common origin in the presence of excess water, develop in different ways and at different speeds according to the climate and geology of the region and according to the geomorphological processes taking place. The buildup of peat is equivalent to the *in situ* storage of much of the production of the ecosystem and depends on the inhibition of the processes of decomposition, erosion and export of organic materials. The identification of the wetland site as marsh, fen or bog is a first step towards understanding its hydrology and the collection of stratigraphic data can yield valuable insights into its past development and its possible response to change.

The work described in this book has taken place on wetland sites which are remarkable for their long-term records of hydrological variables, notably water levels. In this respect these are not typical sites; at most wetland sites there is little or no past data and the only information about hydrological changes comes from records of the loss of species or even the analysis of old place names. To avoid needless repetition of material from an earlier chapter, the sites are summarised in the following table:

Cors Erddreiniog, Anglesey	Fen site with adjoining grazing land
Cors-y-Llyn, Powys	Basin mire with bog communities
Crymlyn Bog, Swansea	Large complex fen site in industrial surroundings
Llangloffan Fen, Dyfed	Riverside fen affected by river channel deepening
Skew Bridge Bog, Powys	Valley bog selected for trials of lysimetry techniques
West Sedgemoor, Somerset	Drained fen under agricultural use
Wicken Fen, Cambridgeshire	Fen managed to maintain habitat diversity

TABLE 5 Sample sites used in this study

THE WETLAND WATER BALANCE

Central to wetland hydrology is the measurement of water level, particularly the elevation of the water table. The seasonal variation of water levels tells much about the natural water regime and periodic measurements, at intervals of a week or a fortnight, are sufficient to provide a good picture of the response of the site to natural and artificial influences. The importance of accurate survey to the long-term value of the records cannot be over-emphasised.

In general, the pattern of wetland groundwater levels takes the form of a relatively high and constant water level in winter and spring followed by a decline in early summer to a minimum in late summer and a rise during autumn, reaching the winter level around mid-winter. Only high rainfall amounts in summer are sufficient to counteract the drawdown effect of high transpiration rates.

Open water levels are controlled by processes taking place in the groundwater body and in the channels of the drainage network. The maintenance (or lack of it) of channels can have an important influence on open water levels but the influence of lowered open water levels on the groundwater body is attenuated and delayed, though the relationship is often complex.

The storage of water in the saturated zone of the soil, an important component of the wetland water balance, is controlled by the specific yield, a hydraulic property of the soil profile which depends strongly on the soil type, in the case of peat on the degree of compaction and humification, and on the position of the water table.

The measurement of water levels in peatlands is complicated by the phenomenon of ground movement. The ground level rises and falls with the water table and both seasonal changes and long-term trends have been measured. Experiments at West Sedgemoor have demonstrated that the main cause of ground movement is the change in buoyancy forces which support the surface peat when it is saturated but not when it is above the water table. Shrinkage of the dewatered peat above the water table has a less dramatic effect on the ground level.

The water table in wetlands is usually so close to the ground surface that heavy rainfall has an obvious and immediate effect on it. The magnitude of this effect, in response to known rainfall amounts, can be used to give an estimate of the specific yield provided allowance is made for the losses from interception during small rainfall events and the likelihood of surface runoff resulting from large events. Values of the specific yield obtained by this means for Cors-y-Llyn, Llangloffan Fen and Wicken Fen (all sites at which continuous water level recorders have been operated) accord well with estimates derived by other workers for these or similar sites and materials, but the short-term response of the water table retains interesting and puzzling features which would justify further investigation.

One of the most important aspects of wetland hydrology is the evaporation, which causes the dramatic decline in the water table in the summer. There have been many studies of wetland evaporation rates but such is the range of wetland habitats, and of the climates in which they are found, that no definitive answer has been presented to the question of whether wetland communities evaporate at a higher or lower rate than open water. A wide-ranging and closely argued study by Crundwell (1986) presents the case for an enhanced evaporation rate but this is based largely on studies of swamp communities in continental climates.

The diurnal fluctuation of the water table in response to the daily cycle of evaporation offers a relatively simple method for determining the daily evaporation. The water table is drawn down by the demands of plant roots and the diurnal fluctuation can be taken as a measure of transpiration, which in summer is the most important component of total evaporation. This method has been investigated in detail using data from Wicken Fen. When daily determinations of the transpiration are converted to monthly values, a pattern emerges which is closely related to the seasonal pattern of growth of the fen community and to the two-year management cycle. Dead material in the standing crop, a feature of natural and semi-natural vegetation communities, is particularly significant in reducing the transpiration demand below the potential rate.

Open drains in wetlands are often regarded as significant sinks for groundwater, drawing down the water table and creating oxidation and wastage of peat for considerable distances. As with other aspects of wetland hydrology, there are few data sets with which to prove or disprove this assertion. This study has the distinction of bringing together for the first time data from dipwell transects at three sites, Cors Erddreiniog, West Sedgemoor and Wicken Fen. At all three sites the influence of the water level in ditches on groundwater levels in the adjacent peat is shown to be confined to a narrow strip no more than 50 m wide. Against this must be set the probable influence of ditches on surface water, particularly in the winter, which reduces flooding and the prevalence of very high groundwater levels by intersecting the natural and semi-natural network of shallow surface channels, ranging from drainage grips and natural 'water tracks' to the runnels between the tussocks of purple moor-grass (*Molinia caerulea*).

A groundwater model, designed to simulate the response of West Sedgemoor groundwater levels to changing ditch water levels, was used to estimate the rates of lateral groundwater flow towards and away from the ditches over the summer of 1987. Groundwater flow from the ditches during the summer has the effect of reducing the drawdown of groundwater levels. The contribution of the ditch network to the water balance of field areas during the summer is a significant proportion of the total input to the system.

The diurnal fluctuations in the water table can be used to determine the net lateral groundwater flow which results in an overnight rise in the groundwater level, partially compensating for the day-time fall. For a large number of days' records at Cors-y-Llyn, lateral flow was found to be predominantly outward, though slow. This provided independent confirmation of the ombrogenous nature of the mire expanse at Llyn.

STUDYING THE WETLAND WATER BALANCE BY LYSIMETRY

The lysimetric method, which is based on the isolation of a block of soil *in situ* and complete with undisturbed vegetation, offers a refinement over the analysis of diurnal fluctuations, in that the effects of lateral groundwater flow (or surface flow) can be eliminated physically. However, if the lysimeter is to be fully representative there must be close control of water level. In the past this control has been exercised by operators in the field, and the lysimetric method has been quite labour-intensive.

The IH wetland lysimeter takes advantage of modern micro-electronics, by using a microcomputer to take over the functions of the operator, performing pumping tests at night to determine the specific yield and exercising close control of the groundwater level by pumping precisely-measured quantities of water in and out of a calibrated reservoir.

The ultimate output of the wetland lysimeter is an estimate of daily transpiration for days on which the water level record is not disturbed by rainfall. Comparisons with daily estimates of the potential transpiration rate for short grass, for Skew Bridge Bog, Powys, and West Sedgemoor, Somerset, corroborate the results from Wicken Fen, showing that transpiration rates are quite low from the herbaceous mire community, whose standing crop always contains dead material, while transpiration from the grazed and mown grass of West Sedgemoor is almost at the potential rate.

KEY POINTS

None of the interpretative work contained in this report would have been possible without the dedicated collection of water level data in the field. In the management of any wetland site, water level data must be a very high priority as it affects the assessment of the effectiveness of water management, the evaluation of changes in the habitat and the confidence with which predictions of future behaviour can be made. In these days of cheaper micro-electronics and more expensive staff time, consideration must be given to the replacement of periodically-read dipwells and continuous chart recorders by data loggers, which offer important savings in time both in the field and in the office, but it is doubtful that the manually-read dipwell network can ever be completely replaced.

The influence of drainage ditches remains a significant and controversial issue in wetland conservation. Data from Cors Erddreiniog, West Sedgemoor and Wicken Fen, featured in this book, and data from Llangloffan Fen which were less clear because of rapid spatial changes in soil type and vegetation community, confirm the conclusion of Boelter (1972) that the zone of influence of a ditch on groundwater levels is very limited. However, the layered structure of some mires, in which a zone of high permeability occurs near the surface, can extend the influence of the ditch, and the cutting of drains across a previously undisturbed peat expanse will draw down the water table permanently into the lower layer, the catotelm. Gowing (1977) noted that lowered Lode levels at Wicken Fen may have been behind the disappearance of the effects of lateral groundwater flow near to Drainers' Dyke between the 1930s and the 1970s, and the peat wastage recorded by ground survey at Cors Erddreiniog extended much further than the measured drawdown effect of the Main Drain, suggesting that 19th century drainage had lowered the water table over a wide area. The key to management of wetland areas adjacent to open drains, for example peripheral ditches, may lie in the manipulation of high water levels, for instance the water levels prevailing in winter and spring, rather than in trying to prevent drawdown by ditches in summer. Certainly the wetland manager must ensure that the ditch does not intercept surface water from his site and he must satisfy himself that the winter flooding regime of the site will not be changed by proposed new ditching or channel improvements.

Wetlands offer those concerned with water resources a groundwater reservoir which combines a degree of natural water treatment, the removal of nutrients, organic pollutants and possibly metals from incoming waters, with flow regulation, the attenuation of floods and the maintenance of a steady flow in dry periods. While it has been demonstrated that the beneficial effects of upland peat on flows are less than previously thought, large valley sites are still valued. There is however a question mark over the water use of a large wetland; if this exceeded the water use of improved land, or even worse if it exceeded the potential rate as has been suggested in the literature, the case for conservation of wetlands on water resources grounds would be less convincing. The results presented in this report, from three contrasting wetland communities, the tall fen of Wicken, the intermediate-height sedges and rushes of Skew Bridge Bog and the managed wet grassland of West Sedgemoor, suggest that wetland transpiration does not exceed the potential rate in the UK. In particular, natural communities, in which the crop is not rejuvenated by cutting or heavy grazing, appear to have quite low transpiration rates, and it must be concluded from the work described in this book that wetlands may have a positive contribution to make to water resources in many circumstances.

Acknowledgements

There have been many contributors to the work contained in this study, over a long period. The author wishes to express his thanks and appreciation to: MAFF Flood Protection (Inland Waters) Commission A, who provided funding for the work on lysimetry and for the collation, analysis and detailed interpretation of wetland data; Nature Conservancy Council for funding of fieldwork on Cors Erddreiniog, Cors-y-Llyn, Crymlyn Bog, Llangloffan Fen, West Sedgemoor and Wicken Fen; NRA-Wessex for funding of supporting work on West Sedgemoor.

Fieldwork on wetland sites is difficult, wet and uncomfortable, and often involves walking considerable distances over treacherous ground. Special thanks are due to those who have maintained the water level measurement programmes over a number of years: Les Colley (NCC warden) on Cors Erddreiniog during and following the Anglesey Fens project; Mike Cox, Paul Franklin, A. Headley, M.R. Hughes, R. Preece and Bob Sutton (NCC) on Crymlyn Bog; Bob Haycock (NCC) on Llangloffan Fen; Andrew Ferguson (NCC) on Llyn Mire; Elizabeth Long, Julia Dixon and Mike Whiteford (IH Wallingford) assisted by sandwich course students Patrick Jefferson, Gillian Austen and Sophie Wood, on West Sedgemoor; Tim Bennett (National Trust warden) and his staff, particularly M. Bentley and R. Fisher, on Wicken Fen.

Colleagues have been co-opted at various stages in the project and have brought new energy and fresh insights as well as technical knowledge: Mary Turner and her colleagues at IH Wallingford, who devised the micro-electronic hardware for the lysimeter and helped to program it; Graham Leeks (IH Plynlimon), Anne Roberts (IH Wallingford) and sandwich course students Ian Thomas and Huw Edwards who helped with the installation of lysimeters; Malcolm Newson (formerly IH Plynlimon, now of the University of Newcastle) for help with the planning and installation of equipment at Cors Erddreiniog, Crymlyn Bog and Wicken Fen and for valuable discussions and preparatory work at the start of this project.

References

Anderson, M. & Idso, S.B. (1985) Evaporative rates of floating and emergent aquatic vegetation: water hyacinths, water ferns, water lilies and cattails, *17th Conf. on Agric. & Forest Meteor. & 7th Conf. on Biometeorol. & Aerobiol.*, May 21-24, 1985, Scottsdale, Ariz., Am. Met. Soc., Boston, Mass.

Bay, R.R. (1966) Evaluation of an evapotranspirometer for peat bogs, *Wat. Resour. Res.* **2**, 437-442.

Belding, E., Rycroft, D.W. & Trafford, B.D. (1975) The drainage of peat soils, *Tech. Bull. 75/2*, Agric. Dev. & Advis. Serv. Field Drainage Exp. Unit.

Beven, K. & Germann, P. (1980) The role of macropores in the hydrology of field soils, *IH Report No 69*, Wallingford, UK.

Beven, K. & Germann, P. (1981) Water flow in soil macropores — II A combined flow model, *J. Soil. Sci.* **32**, 15-29.

Bilham, E.G. (1938) *The Climate of the British Isles*, Macmillan, London.

Boatman, D.J., Goode, D.A. & Hulme, P.D. (1981) The Silver Flowe — III Pattern development on Long Loch and Craigeazle mires, *J. Ecol.* **69**, 897-918.

Boelter, D.H. (1964) Water storage characteristics of several peats in situ, *Soil Sci. Soc. Am. Proc.* **28**, 433-435.

Boelter, D.H. (1965) Hydraulic properties of peats, *Soil Sci.* **100**, 227-231.

Boelter, D.H. (1969) Physical properties of peats as related to the degree of decomposition, *Soil Sci. Soc. Am. Proc.* **33**, 606-609.

Boelter, D.H. (1972) Water table drawdown round an open ditch in organic soils, *J. Hydrol. 15*, 329-340.

Burgy, R.H. & Pomeroy, C.R. (1958) Interception losses in grassy vegetation, *Trans. AGU* **39**, 1095-1100.

Burke, W. (1961) Drainage investigation on bogland: the effect of drain spacing on ground water levels, *Irish J. Agric. Res.* **1**, 31-34.

Calder, I.R. (1976) The measurement of water losses from a forested area using a 'natural' lysimeter, *J. Hydrol.* **30**, 311-325.

Chapman, S.B. (1965) The ecology of Coom Rigg Moss, Northumberland — III Some water relations of the bog system, *J. Ecol. 53*, 371-384.

Crundwell, M.E. (1986) A review of hydrophyte evapotranspiration, *Rev. Hydrobiol. Trop.* **19**, 215-232.

Darby, H.C. (1956) *The Draining of the Fens*, Cambridge University Press, Cambridge.

Dolan, T.J., Hermann, A.J., Bayley, S.E. & Zoltek, J. (1984) Evapotranspiration of a Florida, USA, freshwater wetland, *J. Hydrol.* **74**, 355-371.

Eidmann, F.E. (1959) Die Interception in Buchen- und Fictenbeständen; Ergebnis mehrjähriger Untersuchungen im Rothaargebirge (Sauerland), *Intern. Un. Geodesy & Geophys. & Intern Ass. Sci. Hydrol. Symp. Hannoversch-Münden, Pub.* **48**, 5-25.

Evans, A.H. (1925) Wicken and Burwell Fens fifty years ago and now. In J.S. Gardiner (Ed.) *The Natural History of Wicken Fen*, Bowes & Bowes, Cambridge, 87-91.

Farren, W.S. (1926) A note on the levels of the fens around Wicken, In J.S. Gardiner (Ed.) *The Natural History of Wicken Fen*, Bowes & Bowes, Cambridge, 190-197.

Germann, P. & Beven, K. (1981a) Water flow in soil macropores — I An experimental approach, *J. Soil Sci.* **32**, 1-13.

Germann, P. & Beven, K. (1981b) Water flow in soil macropores — III A statistical approach, *J. Soil Sci.* **32**, 31-39.

Gilman, K. & Marshall, D.C.W. (1991) The West Sedgemoor hydrological study 1986-1991, Report prepared for Nature Conservancy Council.

Gilman, K., Marshall, D.C.W. & Dixon, A.J. (1990) Hydrological studies of West Sedgemoor 1989-1990, Report prepared for Wessex Region of the National Rivers Authority.

Gilman, K. & Newson, M.D. (1983) The Anglesey wetlands study, *IH Special Report*.

Godwin, H. (1931) Studies in the ecology of Wicken Fen — I The ground water level of the fen, *J. Ecol.* **19**, 449-473.

Godwin, H. (1936) Studies in the ecology of Wicken Fen — III The establishment and development of fen scrub (carr), *J. Ecol.* **24**, 82-116.

Godwin, H. (1940) A Boreal transgression of the sea in Swansea Bay, *New Phytologist* **39**, 308-321.

Godwin, H. (1978) *Fenland: Its Ancient Past and Uncertain Future,* Cambridge University Press, Cambridge.

Godwin, H. & Bharucha, F.R. (1932) Studies in the ecology of Wicken Fen — II The Fen water table and its control of plant communities, *J. Ecol.* **20**, 157-191.

Gosselink, J.G. & Turner, R.E. (1978) The role of hydrology in freshwater wetland ecosystems. In Good, R.E., Whigham, D.F. & Simpson, R.L. (Eds) *Freshwater Wetlands — Ecological Processes and Management Potential*, Academic Press, New York, 63-78.

Gowing, J.W. (1977) The hydrology of Wicken Fen and its influence on the acidity of the soil, Unpublished MSc thesis, Cranfield Inst. of Tech.

Greenly, E. (1919) *The Geology of Anglesey*, Mem. Geol. Surv.

Hanson, M. (1976) Mapping the distribution of shell marl in Wicken Fen, Unpublished undergraduate dissertation, University of Cambridge.

Haslam, S.M. (1970) The development of the annual population in *Phragmites communis* Trin., *Ann. Bot.* **34**, 571-591.

Heikurainen, L. (1963) On using ground water table fluctuations for measuring evapotranspiration, *Acta Forestalia Fennica* **76**, 5-16.

Horton, J.H. & Hawkins, R.H. (1965) Flow path of rain from the soil surface to the water table, *Soil Sci.* **100**, 377-383.

Hutchinson, J.N. (1980) The record of peat wastage in the East Anglian Fenlands at Holme Post, 1848-1978, *J. Ecol.* **68**, 229-249.

Ingram, H.A.P. (1978) Soil layers in mires: function and terminology, *J. Soil Sci.* **29**, 224-227.

Ingram, H.A.P. (1982) Size and shape in raised mire ecosystems: a geophysical model, *Nature* **297**, 300-303.

Ingram, H.A.P. (1983) Hydrology. In A.J.P. Gore (Ed.) *Mires: Swamp, Bog, Fen and Moor - A. General Studies*, 67-158.

Institute of Geological Sciences (1969) 1:50000 scale Geological map (Solid & Drift edition), Sheet 296 — Glastonbury.

Institute of Geological Sciences (1972) One-inch geological map Sheet 247 — Swansea (Drift edition).

Institute of Geological Sciences (1974) 1:50000 Geological map (Solid & Drift edition), Sheet 188 — Cambridge.

Institute of Geological Sciences (1975) 1:50000 scale Geological map (Solid & Drift edition), Sheet 295 — Taunton.

Ivanov, K.E. (1975) *Water Movement in Mirelands*, (tr. Thomson, A. & Ingram, H.A.P.), Published in English 1981 by Academic Press, London.

King, F.H. (1892) Observations and experiments on the fluctuations in the level and rate of movement of ground water on the Wisconsin Agricultural Experiment Station farm, *US Weather Bureau Bull. 5.*

Kirby, C., Newson, M.D. & Gilman, K. (1991) Plynlimon research: the first two decades, *IH Report No. 109.*

Lowe, J.J. & Walker, M.J.C. (1984) *Reconstructing Quaternary Environments*, Longman Scientific & Technical, Harlow.

MAFF (1989) Environmentally sensitive areas — First report under section 18(8) of the 1986 Agriculture Act, HMSO, London.

Marshall, D.C.W. & Gilman, K. (1989) Hydrological studies of West Sedgemoor 1986-1989, Report prepared for Wessex Rivers.

Meade, R. & Blackstock, T.H. (1988) The impact of drainage on the distribution of rich-fen plant communities in two Anglesey basins, *Wetlands* **8,** 150-177.

Moore, P.D. & Beckett, P.J. (1971) Vegetation and development of Llyn, a Welsh mire, *Nature* **231,** 363-365.

Moore, P.D. & Webb, J.A. (1978) *An Illustrated Guide to Pollen Analysis,* Hodder & Stoughton, London.

Nicholson, I.A., Robertson, R.A. & Robinson, M. (1989) The effects of drainage on the hydrology of a peat bog, *Int. Peat J.* **3,** 59-83.

Ratcliffe, D.A. (1977) *A Nature Conservation Review,* Cambridge Univ. Press, Cambridge.

Rowell, T.A. (1983) The history and management of Wicken Fen — Discussion Paper 1: Landuse at Wicken Fen since c1600, Dept. of Appl. Biol., University of Cambridge.

Seddon, B.A. (1957) Diagrams from draft of PhD thesis submitted to University of Cambridge in 1958.

Smid, P. (1975) Evaporation from a reedswamp, *J. Ecol.* **63,** 299-309.

Smith, L.P. (1976) The agricultural climate of England and Wales, *MAFF Technical Bulletin 35,* HMSO, London.

Sparling, J.H. (1966) Studies on the relationship between water movement and water chemistry in mires, *Can. J. Bot.* **44,** 747-758.

Sutherland, P. & Nicholson, A. (1986) *Wetland — Life in the Somerset Levels,* Michael Joseph, London.

Thomas, D.W. (1976) A survey of outstanding peatland sites in Gwynedd, NCC Internal Report.

Walker, D. (1970) Direction and rate in some British post-glacial hydroseres. In Walker, D. & West, R.G. (Eds) *Studies in the Vegetational History of the British Isles,* Cambridge University Press, Cambridge, 117-139.

Wallace, J.S., Roberts, J.M. & Roberts, A.M. (1982) Evaporation from heather moorland in north Yorkshire, England, *Proc. Symp. Hydrol. Res. Basins,* Sonderh. Landeshydrologie, Bern.

Walters, H. & Leith, H. (1960) Klimadiagramm-Weltatlas Jena.

White, W.N. (1932) A method of estimating groundwater supplies based on discharge by plants and evaporation from soil, *US Geol. Surv. Water-Supply Paper 659,* 1-106.

Williams, M. (1970) *The Draining of the Somerset Levels,* Cambridge University Press, Cambridge.